Contents

	Preface	v
	Chronology of Chinese History	vii
	Provinces and Major Cities of China	viii
1.	Chinese Dragon Culture	1
2.	Mountains and Rivers in China	5
3.	Religion	13
4.	Chinese Etiquette	19
5.	Traditional Chinese Festivals	25
6.	Chinese Gardens	33
7.	Chinese Residences	39
8.	Famous Temples in China	45
9.	The Lunar Calendar and the Twenty-four Solar Terms	49
10.	Great Chinese Inventions	55
11.	Chinese Characters	61
12.	The Four Treasures of the Chinese Study	67
13.	Lute, Chess, Calligraphy, and Painting	73
14.	Chinese Medicine	79
15.	The Chinese Way to Stay Healthy	85
16.	Wushu	91
17.	Traditional Chinese Weapons	97
18.	Chinese Names	103

19. The Chinese Zodiac 109
20. Chinese Marriage Customs 115
21. Chinese Money 121
22. Chinese Food Culture 127
23. Chinese Alcohol 135
24. Chinese Tea 139
25. Chinese Clothing 145
26. Chinese Embroidery 151
27. Chinese Furniture 157
28. Traditional Folk Arts 163
29. Chinese Jade 169
30. Chinese Porcelain 173
31. Paper Cutting 179
32. Chinese Seals 185
33. Chinese Musical Instruments 189
34. Peking Opera 195
35. Sichuan Opera: Face Changing 201
36. Acrobatics 205
37. Crosstalk 211
38. Auspicious Culture 215
39. Cultural Taboos 223
40. Chinese Feng shui 229
Image Sources 233
Endnotes 239

Snapshots of Chinese Culture

Zhao Yin & Cai Xinzhi

Snapshots of Chinese Culture

For information contact Bridge21 Publications, LLC, 11111 Santa Monica Boulevard, Suite 220, Los Angeles, California 90025

The Chinese edition is originally published by Science Press in 2011. This edition is published by arrangement with Science Press, Beijing, China.

Printed in the United States of America

Cover design by Chi-Wai Li
Maps and timeline by Chi-Wai Li
Cover images from Bigstock.com

ISBN: 978-1-62643-002-0
Library of Congress Control Number: 2013948777

Bridge21 Publications, LLC 11111 Santa Monica Boulevard, Suite 220, Los Angeles, California 90025

Preface

As the world globalizes, China is playing an increasingly important role in world commerce and trade, political and peacekeeping efforts, and more. Now that China has re-opened its doors to the world, many Chinese people study and work abroad, while people from all over the world visit China. Many foreign entrepreneurs not only actively develop their business in China, but also show great interest in learning about its long history and rich culture. With cross-cultural communication ever increasing, the whole world is facing the great challenges and changes that are intrinsic to cultural differences. Differences in customs, ways of thinking, and social background often result in conflict and can be barriers to communication. In our English-teaching careers, we have encountered many excellent native Chinese English learners who have passed the College English Test level 6, and even the Test for English Majors Band 8 (the highest levels), but cannot introduce Chinese customs, philosophy, and arts to foreigners. Thus, we believe we have a great responsibility to promote Chinese culture and allow China to be better understood by the world.

The purpose of this book is to introduce genuine Chinese culture to English speakers, as well as to provide English-language materials to those who are devoted to promoting Chinese culture. It will also serve as a platform for Chinese who cannot spare the time to systematically study their own culture and customs. In addition, it can be used English learners to improve their English as reading material or a text book, or by students of Chinese language who seek to understand culture through language.

Particular debts are owed to the excellent editors we worked with—especially Mr. Nicholas John O'shen—an American teacher who spent much time and energy in editing and proofreading.

Zhao Yin & Cai Xinzhi

Chronology of Chinese History

Ancient China

2852 BC -2070 BC	3 Sovereigns and 5 Emperors	
2100 BC -1600 BC	Xia Dynasty	
1600 BC -1046 BC	Shang Dynasty	
1045 BC - 256 BC	Zhou Dynasty	
1046 BC - 771 BC	- Western Zhou	
770 BC - 221BC	- Eastern Zhou	
	- Spring & Autumn period (771 BC - 476 BC)	
	- Warring States period (481 BC - 221 BC)	

Imperial China

221 BC - 206 BC	Qin Dynasty	
206 BC - 220 AD	Han Dynasty	
	- Western Han (206 BC - 9 AD)	
	- Xin Dynasty (9 AD - 23 AD)	
	- Eastern Han (25 AD - 220 AD)	
220 AD - 280 AD	Three Kingdoms	
	- Wei, Shu and Wu	
265 AD - 420 AD	Jin Dynasty	
	- Western Jin (265 AD - 316 AD)	16 Kingdoms (304 AD - 439 AD)
	- Eastern Jin (317 AD - 420 AD)	
420 AD - 589 AD	Southern and Northern Dynasties	
581 AD - 618 AD	Sui Dynasty	
618 AD - 907 AD	Tang Dynasty	
690 AD - 705 AD	- Second Zhou	
907 AD - 960 AD	5 Dynasties and 10 Kingdoms	Liao Dynasty (907 AD - 1125 AD)
960 AD - 1279 AD	Song Dynasty	
	- Northern Song (960 AD - 1127 AD)	Western Xia (1038 AD - 1227 AD)
	- Southern Song (1127 AD - 1279 AD)	Jin (1115 AD - 1234 AD)
1271 AD - 1368 AD	Yuan Dynasty	
1368 AD - 1644 AD	Ming Dynasty	
1644 AD - 1911 AD	Qing Dynasty	

Provinces and Major Cities of China

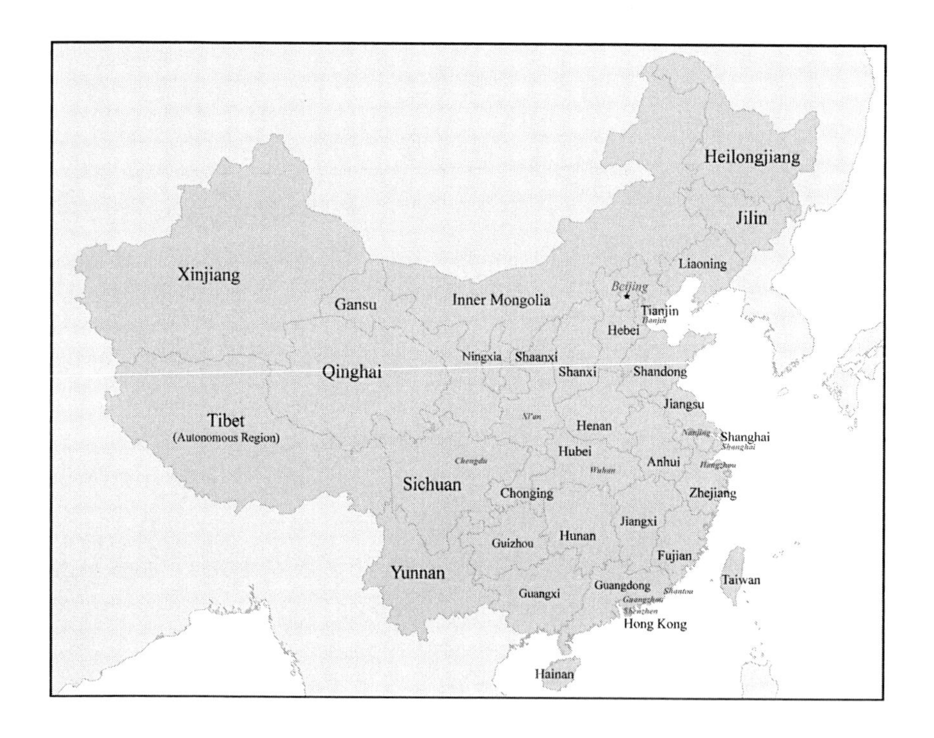

CHAPTER 1

Chinese Dragon Culture

龙文化

The ancestors of the Chinese people, like people of other nations, experienced a clan commune period. At that time, approximately 4,000 BCE, legendary emperors Yandi and Huangdi were leaders of the two most famous clans, belonging to the Huaxia, a confederation of tribes. Because Yandi and Huangdi played an important role in the tribal alliance period, all later generations were considered their descendants. They also called themselves the descendants of the dragon, and Chinese dragon culture continues today.

Many Chinese people like to sing a song: "There is a dragon in the ancient Orient whose name is China. There is a group of ancient Oriental people there too, who are all the descendants of the dragon…" Here, "dragon" is the symbol of the Chinese nation. Some non-Chinese people call China the Asian dragon. Chinese people also like to call themselves the "descendants of the dragon."

In fact, a dragon is a kind of mythological animal in legends. No one has ever seen a flying dragon winding through the clouds in China or anywhere else. It is the mysterious dragon worship, especially the worship of the snake totem, that causes Chinese to think of themselves as the descendants of the dragon.

Some researchers believe that dragon worship formed because some tribes that used snakes as totems combined with tribes that took horses, fish, deer,

1.1 Dragons are placed in front of buildings to ward off evil spirits.

camels, turtles, eagles, cattle, and tigers as totems, thus creating new totems.[1] That is the reason for the newer totemism: the dragon has the deer's horns, the camel's head, the tortoise's eyes, the snake's neck, the scales of the fish, the eagle's claws, the tiger's paws, and the ears of cattle. It is said that dragons can fly, swim, gather clouds to make it rain, and change into different forms.

Since the beginning of the Xia Dynasty, the Chinese have worshiped and revered the dragon. The concept of dragons had already infiltrated every part of society. In ancient people's beliefs, the dragon was a symbol of power and gods, so the Chinese emperors were thought of as dragons. In the past, everything related to the Chinese emperor was connected with the dragon; therefore, the emperor was called "the real dragon and the son of heaven." "The dragon's body is not very well" meant the emperor was ill. When the emperor rewarded someone, the person would say "Thank you, dragon."

The imperial chariot was called a dragon's chariot; the emperor's clothes were called dragon robes, and so on.

The Chinese language demonstrates people's love and respect for the dragon. For example, they use "Hidden dragons and crouching tigers" to describe talented or outstanding people who remain obscure to the untrained eye. "Longing to see one's son become a dragon," means hoping that one's son will have a bright future. The person who is full of vim and vigor is called "as valiant as a dragon and as lively as a tiger." "Fighting between a tiger and a dragon" is used to describe the fierce struggle between well-matched opponents. From the above, we can see the great vitality of ancient dragon culture in China.

1.2 Dragons are a popular decorative motif.

Chinese and Western dragon legends are very different. In the West, the mythological dragon was portrayed as an evil and ferocious animal with large wings and that could breathe fire. In China, the dragon is a symbol of dignity and power.

Chinese Culture through Language

画蛇添足

Pinyin: Huà shé tiān zú

Literally: To draw a snake and add feet.

Meaning: To depict any act which ruins the effect by adding a false or superfluous detail.

画龙点睛

Pinyin: Huà lóng diǎn jīng

Literally: To dot the eyeballs while painting a dragon; to bring the painted dragon to life by putting pupils on its eyes.

Meaning: Adding the crucial touch to a work of art that brings it to life, or saying the word or two that clinches the point.

叶公好龙

Pinyin: Yè gōng hào lóng

Literally: Lord Ye's love of the dragon turns into his worst fear.

Meaning: To depict the professed love of what one really fears.

藏龙卧虎

Pinyin: Cáng lóng wò hǔ

Literally: Hidden dragons and crouching tigers.

Meaning: Talented or outstanding people have yet to be discovered or appreciated; concealed talent.

望子成龙

Pinyin: Wàng zǐ chéng lóng

Literally: To long to see one's son become a dragon.

Meaning: To hope for one's children to succeed.

CHAPTER 2

Mountains and Rivers in China

名山大川

China has a vast territory with an extremely varied geographical structure. It covers about 9.6 million square kilometers (5.96 million square miles) and is the third largest country in the world, after the Russian Federation and Canada. There are many great mountains and rivers in China. The most famous mountains are sanshanwuyue, meaning three great mountains and five sacred mountains in China. There are seven major rivers in China.

When it comes to mountains, Chinese people speak of *sanshanwuyue* (*san shan* means three mountains; 山 is the character for *shan*). Initially, the three great mountains referred to the three mountains of the eight immortals: Penglai, Fangzhang, and Ying Chow. These were not real mountains although they are described as having very beautiful scenery. Today, the three great mountains refer to Huangshan in Anhui Province, Lushan in Jiangxi, and Yandangshan in Zhejiang. The five sacred mountains are the Eastern Mountain, Taishan, in Shandong; the Western Mountain, Huashan, in Shannxi; the Northern Mountain, Hengshan, in Shanxi; the Central Mountain, Songshan, in Henan; and the Southern Mountain, Hengshan, in Hunan. It is said that these mountains were all granted additional titles by many emperors in ancient times. Although they are not the highest and the most precipitous, they appear particularly high and steep, for each is located on a plain or in a basin.

2.1 The famous Huangshan mountain shrouded in clouds.

Huangshan is located in the south of Anhui Province, which is famous for its pines, rocks, many clouds, and hot springs. It is a botanical garden with a lot of flowers, ancient trees, rare birds, and animals. Lushan is in Jiujiang City, Jiangxi Province, and is celebrated for its tall and graceful peaks, unpredictable clouds, magical scenery including a natural spring and waterfall, as well as many historic sites. It is a world-renowned summer resort too. Yandangshan is located in Yueqing County, Zhejiang Province. Because of the lake on the top and the dense reeds where the wild geese are fond of staying during their migration, it is called Yandangshan, meaning Wild Goose Mountain. There are many waterfalls there. The most famous are Big Dragon Pond, Small Dragon Pond, and Three Terraced Falls.

Taishan is located in Shandong Province. It has many unique stones, pine trees, cloud variations, and a magnificent natural landscape. It is called "the number one mountain in China." It is also said that, in ancient times, many emperors worshiped there when they ascended the throne or during peacetime.

2.2 Taishan contains over 1,800 stone tablets and inscriptions.

In 1987, Taishan was listed as a world cultural and national heritage site by UNESCO (United Nations Educational, Scientific, and Cultural Organization).

The Western Mountain, Huashan, is 2,200 meters above sea level, in Shanxi Province. It is the highest of the five sacred mountains, and its natural landscape is peculiar and full of rocks, steep cliffs, deep valleys, and dangerous paths. It is known as the most dangerous mountain.

The Southern Mountain, Hengshan, is located in the middle of Hunan Province. Hengshan's natural scenery is very beautiful: its rich forest is green and full of fragrant flowers all year round. It is a sacred site of Buddhism, and there are many temples, such as the magnificent Nanyue Temple, which covers nearly one hundred thousand square meters.

Shanxi's Hengshan or Northern Mountain is famous for Daoism. It is said to be the location where Zhang Guolao, one of the Eight Immortals in Chinese mythology, once lived. There are a lot of cultural relics and historic sites, such as Hanging Temple, Pagoda of Fogong Temple, and Yuanjue Temple Pagoda in which the structure, shape, decoration, sculpture, and painting that play an important part of Chinese architectural history can be seen.

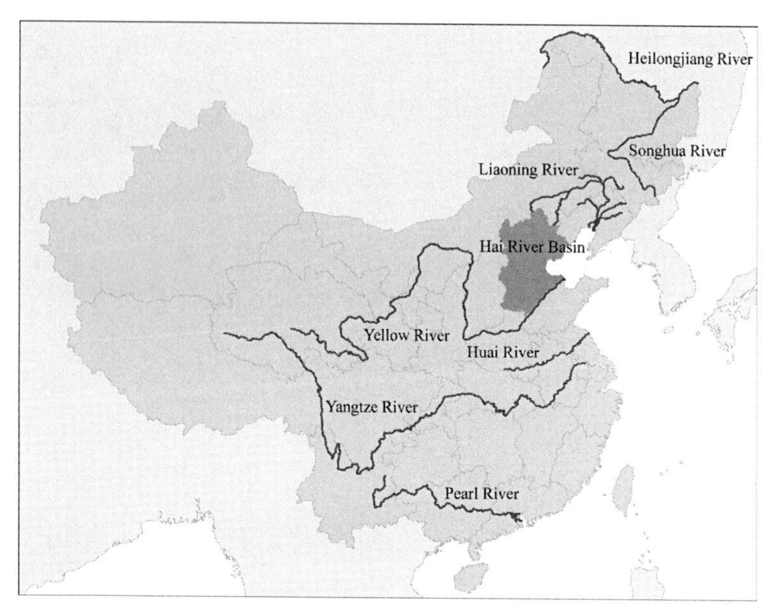

2.3 The largest rivers in China.

The Central Mountain, Songshan, is in Henan Province. Of the thirty-six peaks, Junji is the highest, standing at 1,492 meters above sea level. The Songshan range has many historical monuments, the most famous of which are the Shaolin Temple, Songshan Temple, three carved stones of the Han Dynasty, Zhong Yuemiao Temple, and Huishan Temple.

The seven famous rivers in China are the Yangtze River, Yellow River, Huai River, Pearl River, Hai River, Songhua River, and Liao River. The Yangtze River is the longest in China, 6,300 kilometers. It covers eleven provinces and autonomous regions. It ranks third in the world after the Nile in Africa and the Amazon in South America. The Yellow River, at 5,464 kilometers, is the second largest in China, the cradle of Chinese culture and the mother river of the Chinese people. The Yellow River is a world-famous multi-sediment-laden river. The Huai River is between the Yangtze River and the Yellow River, whose branches permeate four provinces: Henan, Anhui, Jiangsu, and Shandong. The Pearl River is the fourth largest in China at 2,215.8 kilometers, and a small part of it is in Vietnam. The Hai River is the confluence if five rivers, and is the largest river in north China. The Songhua River,

2.4 The Yellow River is considered the cradle of Chinese civilization.

1,927 kilometers long, waters the vast Sanjiang Plain. The 1,430-kilometer-long The Liao River flows past Inner Mongolia and Liaoning Province.

Other relatively well-known rivers are the 2,137-kilometer Tarim River, the largest inland river in China, and Qiantang River, famous for its high tide. The Heilongjiang River is the 10th longest river in the world and runs along the border between Russia and China. China also has the world's longest man-made canal: the Beijing-Hangzhou Grand Canal. It is the largest water conservancy project in ancient China, with the oldest parts dating back to the 5th century BCE. It is 1,764 kilometers long and flows past Beijing, Tianjin, Hebei, Shandong, Jiangsu, Zhejiang, in total six provinces and cities, and links five great rivers: Haihe, Huanghe, Huaihe, Yangtze, and Qiantang.

China's mountains and rivers rival the size of those in the West. The world's highest peak, Mount Everest, is 8,844.43 meters above sea level, and part of it is in China. The longest river is the Nile at 6,695 kilometers. The longest inland river in Europe is the 3,690-kilometer-long Volga River. The longest river in China is the Yangtze at 6,418 kilometers.

The largest freshwater lake is Lake Superior in North America with an area of 82,103 square kilometers. The deepest lake is Lake Baikal, which is 1,940 meters deep. The Suez Canal and the Panama Canal are the best-known canals. All countries have their own famous mountains and great rivers. For example, the Alps are the highest mountain range in Europe, and the highest peak is 4,810 meters. The highest mountain in the United States is Mount McKinley, which is 6,194 meters high.

Chinese Culture through Language

江河日下

Pinyin: Jiāng hé rì xià

Literally: Rivers run downstream on a daily basis.

Meaning: Things or events are getting worse.

愚公移山

Pinyin: Yúgōngyíshān

Literally: Old man Yu Gong moved mountains.

Meaning: The story behind this idiom tells people that so long as one is determined and sticks to one's goal long enough, anything can be done, no matter how difficult.

山盟海誓

Pinyin: Shān méng hǎi shì

Literally: A couple's pledge of love for long-lasting like the mountains and the sea.

Meaning: A solemn, long-lasting pledge of love.

山穷水尽

Pinyin: Shān qióng shuǐ jìn

Literally: Where the hills and streams end.

Meaning: There is no hope left.

CHAPTER 3

Religion
宗教文化

Religion has played an important role in Chinese culture. It was a tool of the ruling class in feudal society and assumed the responsibility of promoting Chinese culture.

China has many religions. According to the China National People's Congress' 2003 Development of Human Rights in China report, there are now more than one hundred thousand religious activities in China, more than three hundred thousand clerics from all religions, more than three thousand national and local religious communities, and more than seventy religious schools. Various religions publish their scriptures, religious books and journals in China, of which twenty million copies are of the Bible.

Early Chinese religious beliefs originated in prehistorical primitive society (before 2070 BC, when the Xia Dynasty was founded). These primitive religions worshipped deities. Based on archaeological discoveries and documents, there were four religious forms popular in China during ancient times. They were nature worship, ancestor worship, totem worship, and reproduction worship. The first included the worship of nature and natural phenomena, such as astrological phenomena. In ancient times, people felt powerless in the face natural phenomena (e.g., wind, rain, thunder, and lightning), so that they worshipped them as celestial beings and were especially in awe of the god of thunder. The second, ancestor worship, came from people's mystical understanding of their own life. They thought that the experience, wisdom, and authority of ancestors would exist within souls forever, which could then act as a supernatural power to protect clan members. From the inheritance of the customs of ancestor worship, the earliest ancestors (the forefathers of all Chinese people) became the symbol of the nation. They then were cherished as gods or divine beings of ancient history. Examples include Yu in the Xia Dynasty, Qi in the Shang Dynasty, and Houji in the Zhou Dynasty; these are all men who became godlike beings. The third religious form was the combination of nature worship and ancestor worship. Each clan had a badge representing its consanguinity. The Chinese goddess Nüwa, credited with creating the earth, is an example: she has the head of a woman and the body of a reptile. Both Emperor Yan and Shennong have been represented with the head of a cow and a human body. The last is the worship of reproduction based on the importance of reproduction to the population of a clan or tribe.

Daoism, an indigenous Chinese religion, was formed in the late Han Dynasty two thousand years ago. It emphasizes simplicity, following the

3.1 A young Daoist monk poses in front of a large incense burner.

lessons of the natural world as they are observed, and *wu wei,* action without acting. Some sects of Daoism believe in immortals and treat immortality as the purpose of life. Taiping Daoism and Five Pecks of Rice Daoism (Grains of Rice Daoist Sect), respectively created by Zhang Jiao and Zhang Daoling, are two branches of Daoism. Daoism integrated folk legend with divination and alchemy, blending the theory of yin and yang and the five elements. It depends primarily on the works of learning by Emperor Huang, which showed the characteristics of Chinese national culture. One of the basic Daoist scriptures is the *Dao De Jing,* composed by the sage Laozi. The Dao, or Way, is considered the origin and master of the universe. Daoist immortality resists the fate of death and the unstoppable cycle of nature. It also directly combats the death of the physical body and hopes the human body could become a spirit. Daoists attempt to possess heaven and earth's forces of creation through human resources and natural and supernatural power. The lifelong work of many Daoists is to be immortal. Daoism incorporates

3.2 Visitors are dwarfed by the giant Buddha of Leshan.

a profound effect on Chinese culture (e.g., medicine, music, architecture, and painting). In the process of its development, Daoist medicine enriched the treasure house of traditional Chinese medicine in both theory and practice. In addition, the four great inventions of ancient China contributed greatly to the scientific development of the world: of the four, gunpowder and the compass are closely related to Daoism.

Chinese Buddhism is a particular form of Buddhism in China. Buddhism was introduced to mainland China from India during the Western and Eastern Han Dynasties, its development period being from the late Eastern Han Dynasty to the Wei, Jin, Southern, and Northern Dynasties. The teachings of the Buddha advocated that all phenomena are illusions. Later, based on the studies of how to reach nirvana (the liberation of the soul, freedom from bodily wants and desires), Buddhism focused on how and why to become a Buddha, and the stages of becoming a Buddha. Chinese Buddhism had a major influence on all classes of feudal society in its long years of development, and continues to influence Chinese culture today. Furthermore, its outward dissemination promoted cultural exchanges among China, its neighboring countries, and the world.

3.3 Guanyin, the goddess of mercy, is revered in both Daoism and Buddhism.

As a vehicle of social culture, Chinese religion played an active role in the development, production, and consolidation of Chinese traditional scientific knowledge.

Similarities and differences exist between Chinese religion and world religions. Each has unique developmental processes, especially Daoism, which historically made great contributions to Chinese science and technology. The discussion of Chinese religion here is in reference to the religions practiced by the Han ethnic group, Daoism, Buddhism, and folk belief.

3.4

Chinese Buddhism not only inherited moral spirit from the religious discipline of Indian Buddhism but also gradually accepted moral values from Confucianism, such as loyalty, filial piety, benevolence, and righteousness. Hence, it enriched Chinese moral norms: the doctrine of piety, the view of compassion, the view of good and evil, and retribution for sin.

Chinese Culture through Language

善有善报，恶有恶报

Pinyin: Shàn yǒu shàn bào, è yǒu è bào

Literally: Do good, reap good; do evil, reap evil.

Meaning: Virtue has its reward, evil its retribution.

半路出家

Pinyin: Bàn lù chū jiā

Literally: To become a Buddhist half way in one's journey.

Meaning: To change to a profession you were not trained for.

不看僧面看佛面

Pinyin: Bù kàn sēng miàn kàn fó miàn

Literally: If not for the monk's sake, do it for the Buddha's.

Meaning: To ask others to do something out of consideration for someone else's reputation.

慧根

Pinyin: Huì gēn

Literally: Wisdom-root.

Meaning: (Buddhism) Talent to thoroughly comprehend Buddhist teachings.

慧眼

Pinyin: Huì yǎn

Literally: Wisdom-eye.

Meaning: (Buddhism) A mind which perceives both past and future; mental discernment; insight; acumen.

禅宗

Pinyin: Chán zōng

Meaning: Zen Buddhism. It can also be called Chinese Zen.

CHAPTER 4

Chinese Etiquette
中国礼仪

China is known as an ancient and a courteous nation. Indeed, Chinese cultural ceremony has influenced and restricted Chinese thought, words, and behaviors profoundly for over five thousand years of historical development. In a sense, the history of Chinese culture is the history of its development of etiquette.

Chinese etiquette originated from the Zhou Dynasty. It was laid down in the *Rites of Zhou,* one of three ancient ritual texts listed among the classics of Confucianism, that social life should follow strict rituals and ceremonies, from sacrifices and deploying troops, to weddings and funerals. The rules of etiquette and folk customs were used to portray hierarchical ranks in society. They showed the difference between superiors and inferiors, kings and subjects, fathers and sons, husbands and wives. Therefore, etiquette and customs reflected most parts of people's life, such as politics, economy, religion, and family. In Chinese traditional society, etiquette has always existed, but the rules varied. This chapter introduces some traditional etiquette in daily life.

Walking Etiquette: In ancient China, people who were in a low position had to bow when they passed someone in a higher position, running with short, quick steps to show respect for seniority. This is called bowing etiquette. In traditional walking etiquette, people should walk along either side of the road but not in the middle of the road, and they should not stand in the middle of doors either.

4.1 A young woman performs gongshou for Chinese New Year.

Meeting Etiquette: In ancient China there were rules for how to meet people of different social ranks. Traditionally, *gongshou* etiquette (also called scraping etiquette) was used for general greetings. To perform the action, make a fist with the right hand and cup it with the left, holding the hands to the chest. In this standing posture, your head and upper body lean slightly forward. This is a greeting to show respect. When visiting someone at his or her house, and especially during holidays, hosts and guests would make this gesture of courtesy and humility when coming into the house and before being seated. This is also scraping etiquette. In addition, people must bow and scrape to elders. Etiquette is used to express thanks, congratulations, apology, or make a request for help. *Gongshou* etiquette came from the Western Zhou Dynasty and is one of China's traditional etiquettes. It also reflects the deep-rooted etiquette of Chinese culture.

Furthermore, in traditional society, people must kneel when meeting the king, kneeling and touching the head and hands to the ground rhythmically. This was called kowtow etiquette and originated from the Eastern Han Dynasty. It is the oldest and the most prevalent social courtesy in Chinese history. Before the Eastern Han Dynasty, people sat on the ground to eat, to do business, and to read. Thus, sitting in ancient times was almost the same as kneeling today. When entertaining guests, the host or hostess would straighten the upper body to show gratitude and respect for the guests (sitting with the back straight). The host then moved from sitting to kneeling, and then bending down (usually touching the forehead to the ground). Gradually, it was called kowtow etiquette. Nowadays it is only practiced in remote villages for New Year, when people greet each other on the first day

4.2 Two civilians bow before government officials.

of the year, traditionally early in the morning. Today, most have adopted the Western practice of shaking hands when they meet.

Seating Etiquette reflects the traditional social order. Seats differ in size and height according to seniority. People of different status have different seating positions. People with higher status sit at the head of the table and with lower status at the foot. Taking a seat improperly could make the host unhappy, is a breach of etiquette, and could be considered insulting and rude. Indoor seats to the east were for distinguished guests, the host's company sitting in the east. Older people would take northern seats facing the south, and the accompanying relatives would take southern seats facing the north. When eating, people are to be as close the table as possible but should sit back when not eating. When distinguished guests visit, the host must stand up immediately to greet them.

Celebrating and Condoling Etiquette: Celebrating etiquette, usually during festivals, was implemented by people of lower status to show reverence to those higher in status and to their peers. People were expected to offer congratulations and bow in worship with a respectful attitude, and to give gifts. Celebrating and condoling etiquette was mainly used during the most important events of life, such as birth, puberty rites, marriage, birthdays, and funerals.

In a word, ancient Chinese etiquette once played an important role in Chinese history and culture. Now, Chinese people have simplified traditional etiquette with characteristics of modern times. Facing the effect of diversification, those who inherit traditional Chinese etiquette and culture are asked to rationally maintain traditions and to keep pace with the development of modern society.

Due to different historical conditions, social ideology, and cultural background, etiquette culture is quite different in China and in Western countries. Bowing in the West is now used primarily for curtain calls and in courtly circles. In the past, some Western people bowed to welcome guests or see them out, to greet ladies, to thank others, etc., though this is no longer popular in modern times. However, Western bowing etiquette does not show the relationship between superiority and inferiority, the old and the young, teachers and students, etc. Nowadays, the most common etiquette both in China and Western countries is to shake hands when meeting others. It has various social functions now and is not only used in adhering to etiquette.

Chinese Culture through Language

礼尚往来，往而不来，非礼也；来而不往，亦非礼也

Pinyin: Lǐ shàng wǎng lái, wǎng ér bù lái, fēi lǐ yě; lái ér bù wǎng, yì fēi lǐ yě

Literally: Courtesy demands reciprocity, courtesy is given but does not not come back, it cannot be maintained. Given and not returned, there is no courtesy.

Meaning: Courtesy demands reciprocity.

君子在德不在衣

Pinyin: Jūn zǐ zài dé bù zài yī

Literally: It's not the gay coat that makes the gentleman.

Meaning: Do not judge a person by his appearance.

一言既出，驷马难追

Pinyin: Yī yán jìchū, sì mǎ nán zhuī

Literally: Once you have said something, the fastest horse cannot get it back.

Meaning: A word spoken cannot be recalled; what has been said cannot be unsaid.

入乡随俗

Pinyin: Rù xiāng suí sú

Literally: Follow the customs once you entered a village.

Meaning: Do as the Romans do.

己所不欲，勿施于人

Pinyin: Jǐ suǒ bù yù, wù shī yú rén

Literally: What you don't want, don't put on others.

Meaning: You should not impose what you don't like to the others.

折柳送别

Pinyin: Zhé liǔ sòng bié

Literally: Gifting a willow branch to someone departing.

Meaning: To bid farewell by snapping off a willow branch, wishing the person departing easily assimilate to local people and customs far away.

CHAPTER 5

Traditional Chinese Festivals

传统节日

China has many ethnicities and a long history. Therefore, there are various traditional festivals, each of which has specific customs, legends, taboos, and celebratory activities. The most important are the Spring Festival, Lantern Festival, Dragon Boat Festival, Mid-Autumn Festival, and Double Nine Festival. During these festivals, people often make some special food to celebrate.

Spring Festival, the traditional Chinese New Year, is the most important traditional festival celebrated by the Han ethnic majority and many of the minority nationalities. It falls on the first day of the first lunar month, commonly known as *guonian* (过年). In fact, Chinese people celebrate the festival from the eighth day of the last month of the lunar year to the fifteenth day of the first month of the following lunar year. In ancient times, festival activities were often related to the land. At the change of the year, people would offer sacrifices from farming and hunting to their ancestors and gods, a token of their gratitude for their bounty as well as a manifestation of their wishes for a bumper harvest in the coming year. On New Year's Eve, people who work and study away from their hometown have to hurry back so the family can have a reunion dinner.

There is a great difference between the reunion dinners of the north and the south. In northern China, people usually eat dumplings or *jiaozi* shaped

5.1 Nuts have auspicious connotations and are offered to family members and guests during holidays.

like a crescent moon or the ancient gold ingot. Some people will put a piece of candy in one of the dumplings to ask for a happy life in the coming year. Many people like to wrap a coin in one dumpling, which means the person who eats it will make a lot of money in the following new year. People in the south eat New Year cakes and *tangyuan,* a kind of round, sweet dumpling. Both the cakes and the sweet dumplings are made of glutinous rice flour. The New Year cake is called *nian gao* in *Putonghua* (the Chinese pronunciation for Mandarin). It conveys the hope of improvement of life year after year. The round, sweet dumpling is a symbol of reunion.

Before and during Spring Festival, many families decorate their houses in the traditional style. People stick the Chinese character *fu* upside down on their door for good luck (see the Chinese Culture Through Language section for more information on this custom), and usually put up couplets of poetry on the front door to bring good fortune. Spring couplets (春联), is a unique literary genre in China, and it is often used during for this decorative purpose. Spring couplets developed from antitheses or antithetical parallelism of *lushi* (a kind of classical Chinese poetry). The couplets must have an equal number of characters, the same rhythm, and the contents must be interconnected. The meaning of the first line must be interrelated to that of the other. A *heng pi* (a horizontal phrase of words above the vertical couplet) often adds the finishing touch to the couplet.

Nowadays, on New Year's Eve, most Chinese people watch the CCTV Spring Festival Evening Show. They also set off firecrackers, and some even stay up all night. On New Year's Day, the whole family pays New Year calls (visits) to the older generation; the children get "lucky money" in red envelopes (*hong bao*) from their parents and grandparents. People of different

5.2 The lion dance is performed to bring good luck and fortune.

regions have their own celebration activities, such as the dragon dance, lion dance, Yangko dance, and others.

Yuanxiao Festival, also called Lantern Festival, is on the fifteenth day of the first Chinese lunar month. In both northern and southern China, family members reunite for a meal of *yuanxiao* (rice dumplings) and enjoy lantern displays. Most of the fillings inside *yuanxiao* are sweet, made of white sugar, brown sugar, sweet-scented osmanthus flower, nuts, and sesame seeds. In recent years, a mixture of Chinese and Western-style *yuanxiao* with chocolate filling has been made. Eating *yuanxiao* is a symbol of the reunion of families in China.

Duanwu Festival falls on the fifth day of the fifth month of the Chinese lunar calendar, also known as Dragon Boat Festival. The main customs are racing dragon boats and eating *zongzi,* filled glutinous rice wrapped in bamboo leaves. It is said that the great patriotic poet Qu Yuan of the Warring States Period threw himself into the Miluo River because of his despair at the crumbling of the State of Chu. The people who loved him threw *zongzi* into the river as food for the fish and shrimp to keep them from eating Qu Yuan's body. *Zongzi* have many flavors and are different in the south and the north. In southern China, people

5.3 *The tradition of dragon boat racing continues around the world.*

make *zongzi* filled with peanuts and pork. Northern people usually use dates or preserved fruit filling.

The Mid-Autumn Festival falls on the fifteenth day of the eighth lunar month. It is said the moon is the fullest and brightest on that day. The ancients made sacrificial offerings of delicious food to the moon, especially moon cake, which people like to eat today. Moon cake filling can be made with pine nuts, walnuts, melon seeds, salted egg yolks, sugar, and other ingredients. Moon cakes are not only delicious but also have beautiful patterns imprinted on them, such as the moon palace and the rabbit and Chang'e flying up to the moon. These patterns remind people into the wonderful myth and story of Chang'e. A Chinese goddess who

5.4 *Traditional moon cakes rest on the caste from which they were made.*

is said to live on the moon. Traditionally, Mid-Autumn Festival is also a day of family reunion.

The Double Nine Festival is a festival for elderly people and is held on the ninth day of the ninth lunar month. As the number nine is *yang* (from the Chinese concept of yin and yang) in ancient China, the ninth day of the ninth lunar month was called Double Yang Festival. Climbing a hill is the most popular activity on that day. It is a custom to drink chrysanthemum wine and Double Nine cakes, also known as chrysanthemum cakes. The cake is made of flour with dates, chestnuts, and meat, like a nine-storied pagoda. Eating Double Nine Festival cakes is symbolic, meaning "be promoted higher and higher every year."

China is a multi-ethnic country, and each ethnicity has its own traditional festivals, such as the Nadam Fair of the Mongolians, Water-splashing Festival of the Dai, Torch Festival of the Yi, Tibetan New Year's Eve, and Third Month Fair of the Bai.

Chinese Culture through Language

每逢佳节倍思亲

Pinyin: Měi féng jiā jié bèi sī qīn

Literally: On festive occasions more than ever, one thinks of dear ones far away.

Meaning: Missing dear ones especially during festivals.

大吉大利！

Pinyin: Dà jí dà lì

Literally: Great luck and great profit!

Meaning: Good luck and great prosperity!

恭喜发财！

Pinyin: Gōng xǐ fā cái

Literally: Congratulations on prosperity!

Meaning: May you come into good fortune!

恭贺新禧！

Pinyin: Gōng hè xīn xǐ

Literally: Compliments to the New Year!

Meaning: Happy New Year!

傣族的泼水节

Pinyin: Dǎi zú de pō shuǐ jié

Meaning: Water-splashing festival of the Dais in southwestern China in April when local folks dress in their holiday best, exchange good will by sprinkling water on each other, and worship the Buddha.

那达慕盛会

Pinyin: Nà dá mù shèng huì

Nadam Fair: a traditional festival of ethnic Mongolians in Inner Mongolia. It is celebrated with events such as wrestling, horse races, archery, and dancing.

福字倒贴

Pinyin: Fú zì dào tiē

Literally: Pasting the Chinese character "fu" up-side down.

Meaning: This is an effort to beckon in happiness or good fortune. When the character for fortune (福, fú) is turned upside down, kids might point to it and say it aloud and that will sound like "fortune has come!".

CHAPTER 6

Chinese Gardens

中国园林

The Chinese garden is an example of wonderful architecture and one of the great treasures of traditional Chinese culture. Chinese gardens, seen as landscaping masterpieces, create natural environments and landscapes with highly spiritual themes and incorporate the theories of feng shui by using flowers, trees, water, and other natural objects and elements.

Chinese gardens have a long history and are famous throughout the world. They are known as one of the world's three major garden systems, along with the gardens of West Asia and Europe. China's rich flora and long history of garden development has had a profound influence on horticulture and garden design throughout the world. Chinese classical gardens represent a harmonious blending of people and nature.

Geographically, ancient Chinese gardens can be divided into two broad categories, the imperial gardens in the north and the private gardens in the south.

The term "imperial garden" generally refers to those in which emperors lived and entertained themselves.[2] They are mainly located in Beijing and the surrounding areas. The design of an imperial garden reflects the idea that the emperor was the richest man in the world. They had a liberal choice of locations, and the construction of the garden was created with abundant resources, so that not only mountains and lakes in their original forms

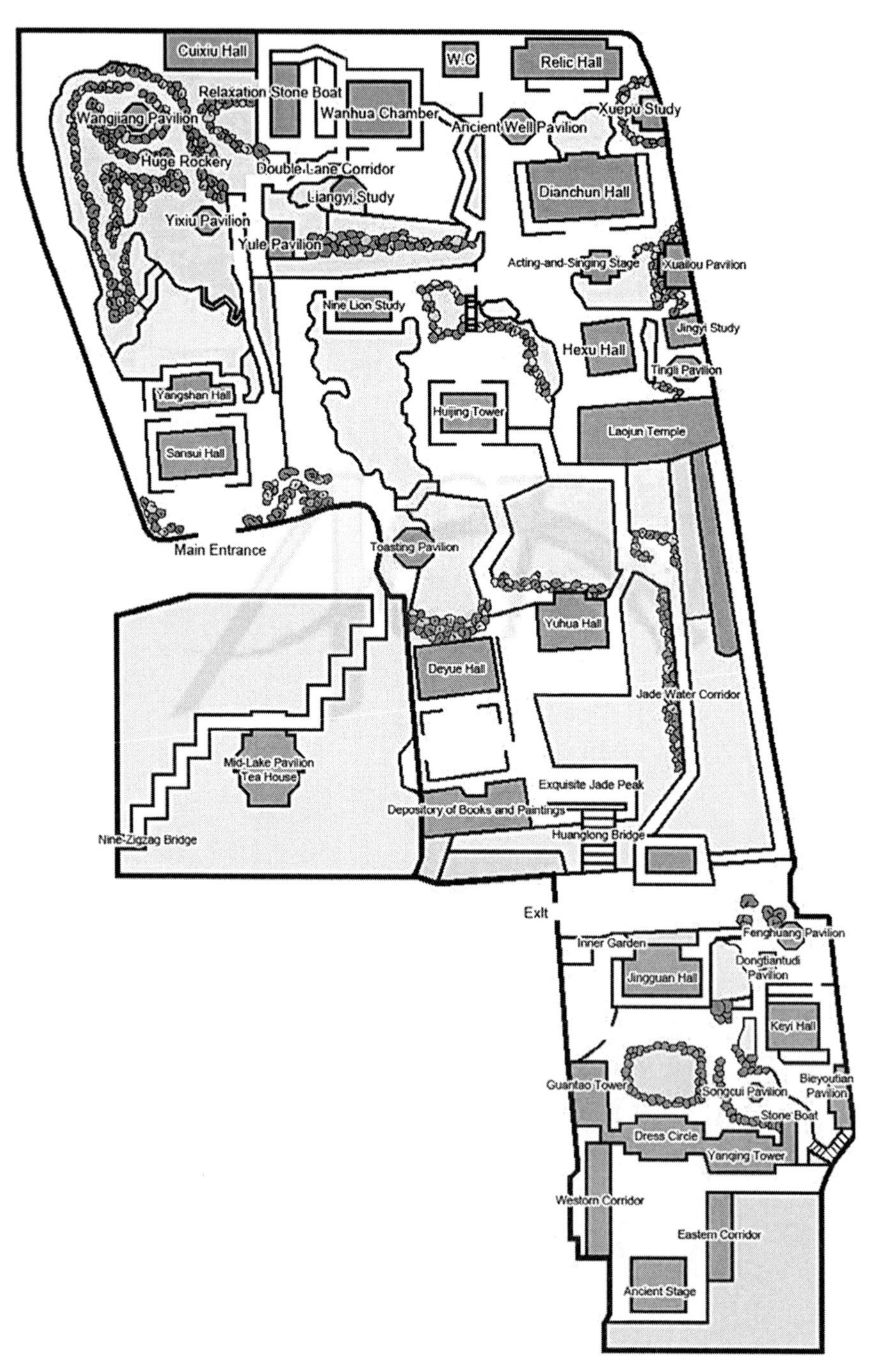

6.1 The Yuyuan Garden of Shanghai was completed in 1577. It occupies an area of 20,000 square meters (5 acres).

were incorporated, but lakes were dug, and mountains constructed, as well, imitating natural scenes. All sorts of buildings were constructed, too.

The general layout of an imperial garden is magnificent: the buildings are beautiful, imposing, and perform diverse functions. A private garden cannot compare with an imperial one.

The famous imperial gardens are Yuanmingyuan, Xiangshan Park, and Beihai Park in Beijing, and the Mountain Summer Resort of Chengde, Hebei; and those inside the Forbidden City. Yuanmingyuan (the Old Summer Palace, originally called the Imperial Garden), known as the Garden of Gardens, has combined different styles of Chinese landscape art and integrated some Western architectural styles. The park is laid out with various intricate patterns of great interest. Yuanmingyuan covers an area of five thousand acres, together with Changchun Park and Wanchun Park. In the garden, there are three major lakes, nine islands and nearly one hundred buildings, which contain superb artistry fit for a large imperial garden. Chengde Summer Resort, which has 8,500 acres, 72 scenic spots and 11 temples, is the largest imperial palace in the area and the largest garden in China.

The private gardens were built for the families of the royal imperial wives, the princes and high officials, and rich merchants. These gardens were created mainly in Shanghai, Jiangsu, and Zhejiang, of which Suzhou's are the most famous and classic. Gardening in Suzhou reached its height during the Ming and Qing Dynasties. Suzhou became the center of a garden supply industry; thus, the gardens of Suzhou have been considered the standard of classical design.

Most of Suzhou's gardens are relatively small but make good use of various forms and charming artistic features with flowers and birds in the limited space. They are decorated by artificial rocks, trees, arranged pavilions, ponds, and bridges, all in exquisite and elegant colors, creating an artistic and poetic effect mimicking the poetry of the Tang and Song Dynasties. There are over 280 private gardens in Suzhou; 69 in and around Suzhou are preserved as important national cultural heritage sites, and some of them are on the World Heritage List.

The Humble Administrator's Garden, Lingering Garden, Blue Wave Pavilion, and Lion Grove Garden are referred to as the four famous gardens in Suzhou. The Humble Administrator's Garden has thirty-one scenic spots within an area of sixty acres. It is water-focused, as about three-fifths of the surface is covered with

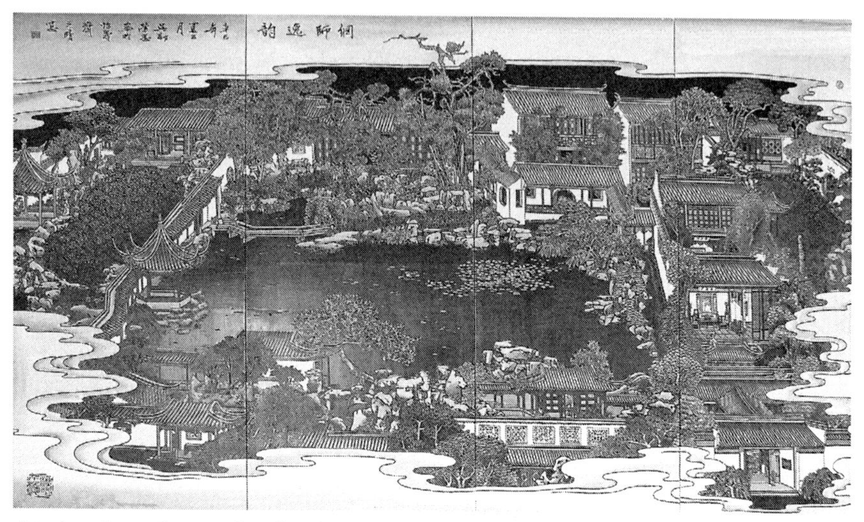

6.2 A traditional watercolor of Suzhou's Master of the Nets Garden.

lakes surrounding building complexes. The whole garden is divided into three sections: east, west, and center. After the founding of the country in 1949 the east entrance was rebuilt. The Laxuetang, Hibiscus Pavilion, Tianquan Pavilion, and Fangyan Kiosks have also been rebuilt (1952–1961). In the middle there is a pool with rocks, trees, and buildings, gracefully and naturally surrounding it. Jinxiang Hall has doors or windows in all directions so that you cannot miss any of the beautiful surrounding spots. The rockery is slightly undulating, facing Shuiyue Platform in the north. Trees such as willows are planted along the pond, and the north gallery has views of Hill House with two storeys surrounded by water on three sides. On the west side is a pile of rocks that people can see winding around a mountain frame. Nanxuan and Xiangzhou are on either side of the pool.

The Humble Administrator's Garden owes its greatness to its many perfectly distributed pools. The layout appears natural by making clever use of the natural environment. It has the typical characteristics of southern gardens and is a superb national cultural relic.

Most important, picturesque Chinese gardens are integrated with painting, literature, gardening, and other forms of art and traditional Chinese culture. This art has not only omitted the manicured features of flowerbeds, buildings, and trees in Western gardens but also the great attention to detail characteristic of the Japanese garden. Chinese gardens are a reflection of nature with artistic skill, attaining implicit perfection. The Chinese garden is unique in its valuable historical and cultural heritage and its contribution to humanity.

The difference between Chinese and Western gardens are the following. Western gardens emphasize mathematical principles of geometry for the main designs, whereas Chinese gardens are based on the natural beauty of the landscape and the viewers' experience of the scenery, focusing more on the harmony between nature and people.

The Summer Palace was one of the most outstanding palaces of its time. Through letters and reports, Western missionaries introduced it, and it became famous in Europe and played a very important part in the development of natural landscape gardens and horticulture in eighteenth-century Europe.

The main characteristics of Chinese gardens are that they are created according to local conditions, pools are dug and mountains created, and the structures are designed in alignment with the flowers, trees, and overall environment to create a natural feeling. In style, Chinese gardens are subtle, elegant and simple, innocent, and delicate; wonderful work is imbedded in them, creating a unique system of the art of Chinese gardens different from those of Western countries, particularly Britain and France.

6.3 Pagodas are a traditional structural element of Chinese garden design.

Chinese Culture through Language

城市绿化

Pinyin: Chéng shì lǜ huà

Literally: City greening.

Meaning: Urban greening, urban planting.

历史名园

Pinyin: Lì shǐ míng yuán

Literally: Famous historical garden(s).

Meaning: An historical Chinese garden or park.

纪念公园

Pinyin: Jì niàn gōng yuán

Literally: Memorial park.

Meaning: A park in memory of historical events and famous people.

居住绿地

Pinyin: Jū zhù lǜ dì

Literally: Green space in living space.

Meaning: Green space attached to housing estates; residential green space.

道路绿地

Pinyin: Dào lù lǜ dì

Literally: Green land along roads.

Meaning: Green space on the road side.

屋顶花园

Pinyin: Wū dǐng huā yuán

Meaning: Rooftop gardens.

CHAPTER 7

Chinese Residences

中国民居

Residential buildings throughout China are the most basic type of building. The environmental and cultural differences between regions have led to residences very diverse in design and construction. Chinese residences are treasures of Chinese traditional architectural history, and they have become an important transmitter of Chinese culture.

Traditional houses refer to houses in villages that are not constructed by the government, are folk-like, and are the private residences that have been inherited from generation to generation. These residences were not designed by architects or craftspeople, which is why they play such a distinctive role in Chinese architectural history. They have unique architectural forms and characterize the socio-economic and cultural development of their regions and environments. The most famous representatives are Beijing's *siheyuan*, Fujian's Hakka group houses, cave dwellings in northern Shaanxi, and *ganlan* in Guangxi.

Beijing's *siheyuan* (courtyard house) are the most important example of Chinese traditional residential housing. Originating in the Yuan Dynasty and popular in Beijing, there are many of them. *Sihe* refers to the east, west, south, north; houses are built on each side to form a mouth-shaped courtyard.

7.1 The Siheyuan structure has over 3,000 years of history.

Siheyuan can often be divided into two halls. The center of the main structure is where family rituals are held and distinguished guests are received, and all the houses are connected by verandas and face the central garden. In general, a formal courtyard house faces south and is surrounded by houses known as the north or principal room, and the south or inverted seat room. There are the east and west wing rooms, and the gate is located in the southeast corner. The open space in the middle, the courtyard, has flowers and trees planted in a way so that the people living in the courtyard house are very close to nature. This fits well with the traditional Chinese cultural concept of Heaven and Human (the theory that man is an integral part of nature, and must also be in harmony with nature).

Siheyuan in Beijing reached their apogee during the Qing Dynasty and had a great influence on the residences of the time and those that followed. These residences became the Han style of architecture.

Lingnan Hakka group houses or Hakka *kakoiya* are traditional dwellings for Hakka in west Fujian, Guangdong, and Guangxi Provinces. They use rammed earth for load-bearing walls, hence the name *tulou* (earth building). There are three to four floors on average, and the tallest house can have up to six floors. Group houses are round or square with a storage house for grain and a kitchen or utility room on the ground floor. People live on the higher floors. The central hall is for family events such as weddings, funerals, rituals,

7.2 Tou lous are quickly being destroyed and replaced by modern buildings.

and other activities. The Hakka generally like to have a pond in front of their houses, which is an important symbol of the Hakka *kakoiya*. Including the apartments within the structure, *tulou* can usually hold more than fifty families. Halls, storage houses, domestic animal houses, wells, and other public houses are all located in the yard. The Hakka employ this special defensive layout to protect themselves, and it is still in use. Among Chinese traditional residences, the Yongding Hakka *tulou* are unique. There are square, circular, octagonal and oval shapes. The total number of housing groups is over eight thousand, and they constitute a wonderful residential world. Their beautiful shapes, features, scientific, and practical designs are truly unique.

Ganlan are mainly distributed in the southwest provinces of China, such as Yunnan, Guizhou, Guangdong, and Guangxi. These are the residences for Dai, Jingpo, Zhuang, and other minority groups. *Ganlan* are wood or bamboo storied houses, supported by poles and usually standing alone, separated from other *ganlan* houses. The living quarters of *ganlan* are usually on the second floor high above the ground, and the first story is retained for raising domestic animals and storage. The entire structure of *ganlan* is very sensible and practical. *Ganlan* can ward off moisture, as well as attacks from insects, snakes, and other animals. *Ganlan* are suitable for people living in the damp and rainy mountains in southern China, often marked by uneven terrain. *Ganlan* can still be seen in parts of China today. For example, in the northern part of Guangxi, layers of terraced fields are dotted by relaxing and tranquil *ganlan* villages.

Cave dwellings, a kind of traditional residence with a long history, are mainly distributed in central and western provinces like Shaanxi, Gansu, and Qinghai, where loess is very deep, strong, and solid enough for cave dwellings. Loess is a product of the accumulation of wind-blown silt and clay into geological formations. It has little seepage and provides very good conditions for cave dwellings. From the relationship between the cave and the ground, cave dwellings can be divided into three types: cliff, ground, and hoop dwellings.

The cliff cave dwelling is an earth cave dug horizontally along the vertical earth cliff, the top dug into a semicircle or circular arch shape to let sun shine in.

7.3 A traditional cave dwelling in Luoyang, Henan.

Ground or pit cave dwellings are dug into square or rectangular pits from the flat ground and have a courtyard. From the inside walls of the pits, caves are dug. Ground cave dwellings appeared in areas where there are no cliffs. From the ground, one can only see treetops and the yard but not caves, such the ones in Luochuan. The hoop cave dwelling (or indium) is not a real cave. It is a cave-shaped house built on the ground with stones, bricks, or adobe. If the upper so-called cave is also a hoop, it is referred to as a cave upon a cave; if the top is a wooden structure, then it is called a house upon a cave. The cave dwelling is cool in summer and warm in winter, and saves space. It is a harmonious combination of natural environment and human activities.

There are also many other special residential houses, such as Jingsu residences, Huizhou residences, Mongolian yurts, Tibetan residences, and even houseboats. The wide variety of forms and characteristics of Chinese residences reveal the complex interactions and factors that affect the various ethnic traditional styles, customs, and traditional culture.

The domestic dwelling is the earliest type of architecture and is the most widely constructed. Dwellings are not only a creation of local environments and lifestyles but also have special and ethnic characteristics. The courtyard in Chinese traditional dwellings contains elements of the Chinese traditional cultural lifestyle from two thousand years ago and has important cultural and social value. Chinese residences have rich cultural connotations too. The great difference from Western dwellings is that the Western dwellings are built on a certain point, whereas the Chinese-style dwellings are built around a certain point. For example, inside a Chinese residence there must first be a courtyard, and all rooms are arranged around it: it is the traditional center of life. If there were not a yard, one may feel it was not a house. There is an external wall which is mostly solid and relatively closed. In contrast, many Western-style houses are compact and close together, open, and built with a garden outside, though this can vary greatly from country to country and between urban and rural settings. Modern Chinese residences are designed to take modern living into consideration and incorporate international design elements. They incorporate many techniques and styles of Western architecture.

Chinese Culture through Language

百万买宅，千万买邻

Pinyin: Bǎi wàn mǎi zhái, qiān wàn mǎi lín

Literally: A million dollars for a house; Ten million dollars for the neighborhood.

Meaning: Neighbors are important, so they must be chosen carefully.

独门独院

Pinyin: Dú mén dú yuàn

Literally: independent-gate-independent-yard.

Meaning: A single housing unit occupied by one family, and is away from the other houses.

安家立业

Pinyin: Ān jiā lì yè

Literally: set up-family-establish-career.

Meaning: Settle down and embark on a career.

三顾茅庐

Pinyin: Sān gù máo lú

Literally: Making three calls at the thatched cottage.

Meaning: To call on someone repeatedly, to repeatedly ask someone to take up an important post.

CHAPTER 8

Famous Temples in China
中国名刹

Chinese religious culture has a long history, Buddhism having the greatest influence. It not only has had a major influence on traditional Chinese thinking but also has left behind a rich architectural and artistic heritage. Daoist temples are scattered throughout China, but Buddhist temples predominate and can be seen both in densely populated cities and scenic mountain landscapes.

Many Chinese temples were built in the mountains. Most were courtyard-style buildings. The main architectural components are the temple door, the Heavenly King Hall, Great Buddha's Hall, the Hall for Preaching Buddhist Doctrines, and the Depository of Buddhist texts. The Maitreya Buddha (known as the laughing Buddha, the next Buddha to be incarnated on earth) is enshrined in the Heavenly King Hall. Four heavenly kings are at side halls, and each protects one direction. They also function to bring the necessary wind and rain for a good harvest. The main hall is a place to worship Shakyamuni, the founder of Buddhism. The east side hall is dedicated to the temple guardian, and the west worships the first ancestor of Zen, Damo.

The hall where the monks read and expound on Buddhist texts is behind the main hall. The Depository of Buddhist texts, the library of the temple, is often built in the inner courtyard.

8.1 The main entrance to Baima Temple.

Arhat Hall is purely Chinese, and there are usually five hundred different lifelike statuaries of *arhats* (beings that have attained nirvana) in it.

White Horse Temple (Baima Temple) in Luoyang is the first temple built by the Chinese government in the year 68. Originally, it was only the dwelling place of monks from other countries. Baima Temple is looked upon as the birthplace of Buddhism in China.

Many temples are located in China's famous mountains. Shaolin Temple is the most famous temple in China and is located in Songshan, Henan Province. During the Tang Dynasty, Shaolin monks rescued the king, Li Shi Min, and won great recognition. Since then the temple has become world famous. In the biggest hall, Qianfodian, there are a lot of shallow pits in the floor, which are said to be the result of the monks' kung fu training. Many martial arts enthusiasts in China and abroad visit Shaolin Temple every year.

Foguang Temple, also known as the gem of ancient Chinese architecture, is in Wutai County, Shanxi Province. The temple stands high on the mountainside with green pines, in a quiet environment. Many visitors visit the temple and admire the charming scenery every year.

Located to the west of the West Lake, Lingyin Temple is the largest and oldest in Hangzhou, having over 1,600 years of history. Lingyinsi Temple is famous for its beautiful Feilai Peak, dignified pavilions, clear cold spring, and the inscriptions carved in the Tang and Song Dynasties. It is a museum of nature, history, culture, and architecture.

8.2 Many foreigners and Chinese visit the Shaolin Temple to practice martial arts.

The largest Lama Temple (lama is a teacher of Tibetan Buddhism) in China is the Potala Palace located in the northwest of Abdallah Hill, Lhasa. It is composed of five different style palaces covered with golden glazed roof tile and magnificent walls that were built with red and white stones. In order to marry Princess Wen Cheng of the Tang Dynasty, the king of Tibet, Songtsan Gampo, built the Potala Palace in the seventh century. Inside, there are more than two hundred thousand statues of Buddha and many fine murals. On the top, there is a row of tall Nepalese-style tomb towers. One of them is of the thirteenth Dalai Lama, which is fourteen meters tall and covered with gold. According to statistics, this statue is composed of 590 kg of gold, 200,000 pearls, and thousands of other kinds of precious stones; thus the temple is known as the most beautiful tomb tower.

There are many other famous temples in China, such as Yonghegong, Kaiyuansi Temple, and Pujisi Temple.

Chinese Culture through Language

普度众生

Pinyin: Pǔ dù zhong shēng

Literally: salvation of all living things.

Meaning: A Buddhist term, meaning to deliver all living creatures from torment.

山门

Pinyin: Shān mén

Literally: Mountain door.

Meaning: The gate of a temple is called the door of the mountain.

跑了和尚跑不了庙

Pinyin: Pǎo le hé shàng pǎo bù liǎo miào

Literally: The monk may run away, but not the temple.

Meaning: A criminal can run away, but will be caught sooner or later.

平时不烧香，急来抱佛脚

Pinyin: Píng shí bù shāo xiāng, jí lái bào fó jiǎo

Literally: To never burn incense when all is well but then clasp the Buddha's feet when in distress.

Meaning: To do nothing until the last minute.

丈二和尚，摸不着头脑

Pinyin: Zhàng èr hé shàng, mō bù zháo tóu nǎo

Literally: You can't touch the head of the ten-foot monk.

Meaning: You can't make heads or tails of something.

放下屠刀，立地成佛

Pinyin: Fàng xià tú dāo, lì dì chéng fó

Literally: Drop the butcher's knife and become a Buddha.

Meaning: A person can achieve salvation as soon as he gives up evil.

CHAPTER 9

The Lunar Calendar and the Twenty-four Solar Terms

中国历法和二十四节气

The creation of the lunar calendar and the solar terms within it is closely related to human life. Because of the development of agriculture, it became more important to understand weather and the solar terms. The ancient Chinese people have created the Chinese lunar calendar and the twenty-four solar terms from observing astronomical phenomena and the climate.

There have been more than one hundred kinds of calendars in Chinese history. At present, only the lunar calendar and solar calendar are commonly used. The solar calendar is based on the rotation of the sun and the four seasons. There are twelve months a year, 365 days, and there is a leap year every four years. The leap year has 366 days; an extra day occurs at the end of February.

The Chinese lunar calendar is aligned with agricultural production. The lunar calendar is based on the cycles of the moon. There are twelve months a year, 354 days; there are thirteen months in a leap year, and the extra month is called the leap month. About seven leap months occur in every nineteen lunar years, and the leap months are based on the twenty-four solar terms of the year. (Solar terms are used in the East Asian lunisolar calendar—a calendar which indicates both the moon phase and the time of the solar year—and

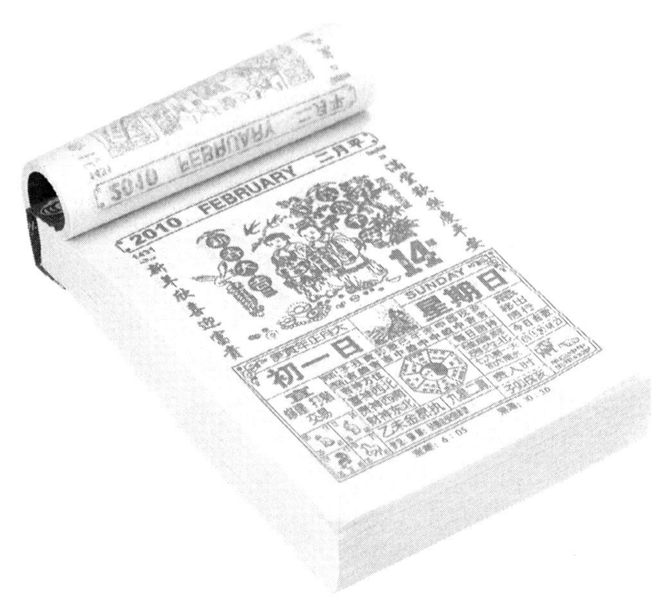

9.1 Traditional Chinese lunar calendar.

keep people in synch with the seasons). The twenty-four solar terms can be divided into twelve major and twelve minor solar terms. Generally, a month without a major solar term more than every two years is called a leap month.

The major and minor solar terms are collectively called solar terms. From the major solar term "moderate cold," there is a minor solar term at approximately thirty-day intervals, such as "spring commencement," "insects waken," "bright and clear," and so on. Starting from winter solstice, there is a major solar term at about thirty-day intervals, with names such as "severe cold," "spring showers," vernal equinox, and so on. Every solar term is about 15.22 days on average.

A Chinese lunar year is divided into four seasons: spring, summer, autumn, and winter, and there are minor solar terms called spring, summer, autumn, and winter commencement within the twenty-four solar terms. Because the days of a solar term are fixed and are not related to the fixed months, the occurrence of the solar terms and the lunar calendar months are not always consistent every year. However, the twenty-four solar terms have laid the foundation for China's farming schedule.

In fact, every solar term has a clear meaning; for example, "great heat" refers to the period when hot weather has reached its peak; "severe cold" means the coldest days of the year have come; "insects waken" indicates that the spring thunderstorms have awakened the hibernating insects and animals; "corn rain" suggests the rainfall will increase, which is beneficial to the growth of corn; "corn on the ear" is the harvest time of wheat, denoting summer planting will begin.

Chinese farmers have created many farming sayings according to the twenty-four solar terms. For example, "Three days after the term of spring commencement, all the plants sprout"; "Spring plowing should not stop after the insects waken term"; "It is not time to plant grain after the term of corn on the ear"; and "It is a busy time during the terms of the autumn commencement and the end of heat, and people have to gather in the early rice and plant the coarse cereals then." The twenty-four solar terms are part of the unique cultural heritage of the Chinese working people and not only

9.2 Lunar calendars come in many forms. Traditionally characters were drawn from top to bottom, right to left.

guide farming activities but also reflect the changes in season. They have been applied by Chinese farmers for thousands of years.

As Chinese people use solar terms for farming, so did the early people of the first American colonies: *Poor Richard's Almanac,* written by the American inventor and statesman Benjamin Franklin, was popular for its weather forecasts as well in addition to its puzzles and stories. There are still many farming cultures around the world which refer to solar and lunar almanacs.

Chinese Culture through Language

立春雨淋淋，阴阴湿湿到清明

Pinyin: Lì chūn yǔ lín lín, yīn yīn shī shī dào qīng míng

Literally: Wet weather will continue until Qing Ming (Pure Brightness) Day if it rains on the Beginning of Spring Day.

Meaning: If it drizzles on the Beginning of Spring Day, the rainy period will last until Qing Ming Day.

小暑热得透，大暑凉飕飕

Pinyin: Xiǎo shǔ rè de tòu, dà shǔ liáng sōu sōu

Meaning: If it is extremely hot on Slight Heat Day (or Mid-Heat Day), it will be very cool on Great Heat Day (Scorching-Heat Day).

雪兆丰年

Pinyin: Xuě zhào fēng nián

Literally: snow foretells a bumper harvest.

Meaning: Heavy snow in winter will assure a good harvest next year.

一年之计在于春，一日之计在于晨，一生之计在于勤

Pinyin: Yī nián zhī jì zài yú chūn, yī rì zhī jì zài yú chén, yī shēng zhī jì zài yú qín

Literally: The whole year's work depends on a good start in spring; the whole day's work depends on a good start in the morning; the success of one's lifework depends on one's industriousness.

Meaning: Plan ahead, work hard.

去年的皇历，今年看不得

Pinyin: Qù nián de huáng li, jīn nián kàn bù dé

Literally: Last year's almanac is useless for this year.

Meaning: What was correct may not be the same case at present.

老皇历

Pinyin: Lǎo huáng li

Literally: old almanac.

Meaning: Ancient history or outdated ways of doing things.

CHAPTER 10

Great Chinese Inventions

中国四大发明

American scholar Robert Temple declared in his book The Genius of China: 3,000 years of Science, Discovery, & Invention that more than half the world's basic inventions are from China.[3] Particularly, the inventions of papermaking, gunpowder, printing, and the compass have changed the world. The science and technology within the country have enabled China to maintain a leading position in the world for a millennium.

Four great inventions of ancient China were revolutionary for the world's development: papermaking, gunpowder, printing, and the compass. They propelled humanity's political, economic, and cultural development forward and created enormous impetus for the development of world civilization.

Paper: The first writing materials were turtle shell, animal bones, bamboo, bark, stone, and other objects. Writing on these was sometimes inconvenient, they were not easy to carry, some were expensive, and they were not suitable for wide use. In 105 BCE, Cai Lun improved papermaking, based on the existing papermaking technology. The principles of ancient paper craft had been developed over two thousand years, but there was no substantive change in the process of paper making for a long time. The basic process went as follows: collect the main raw material (hemp plants); mash, steam, whiten, complete the natural fermentation process, make it into a pulp; form into sheets (with bamboo screens); dry in the sun. Owing to the widespread use of paper, the recording and dissemination of information made revolutionary progress.

10.1 Firecrackers make festivals and weddings lively, and ward away evil spirits.

Gunpowder: As early as the Shang and Zhou Dynasties, people used charcoal widely in metallurgical practice. Furthermore, the raw materials of gunpowder—sulfur and saltpeter—were well known to the Chinese people, thus paving the way for the invention of gunpowder. In the Tang Dynasty, the alchemists gathered the various experiences in production and life from working people. Then they gradually found that a mixture of sulfur, charcoal and other materials could be flammable and explosive. The earliest record of this can be found in Sun Simiao's book *The Method of Using Sulfur*. This book provides evidence that gunpowder was invented in China in the early Tang Dynasty. In 904 CE, during the Tang Dynasty, Zheng Pan made gunpowder to attack Yuzhang. This was the beginning of gunpowder for military use. During the Northern Song Dynasty, gunpowder was widely used in the military and then spread to Arab countries. In the latter half of the thirteenth century, Europeans gradually learned how to manufacture and use gunpowder.

Printing has been recognized as one of the greatest inventions in the world. Before the invention of printing, people used to copy by hand, which was both time-consuming and labor-intensive; moreover, it was prone to errors. In the Sui and Tang Dynasties, people gained inspiration from seal and rock carving. They carved convex characters in wood, brushed ink on them, and covered the characters with paper to create an early form of printing. Nevertheless, the invention of woodblock printing in early Tang Dynasty was a leap forward. It was still a

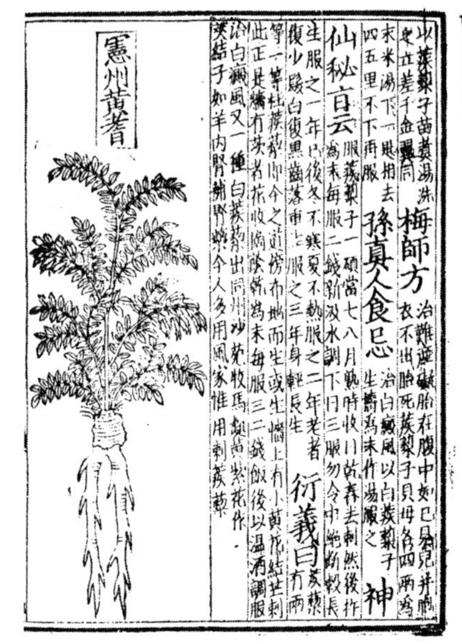

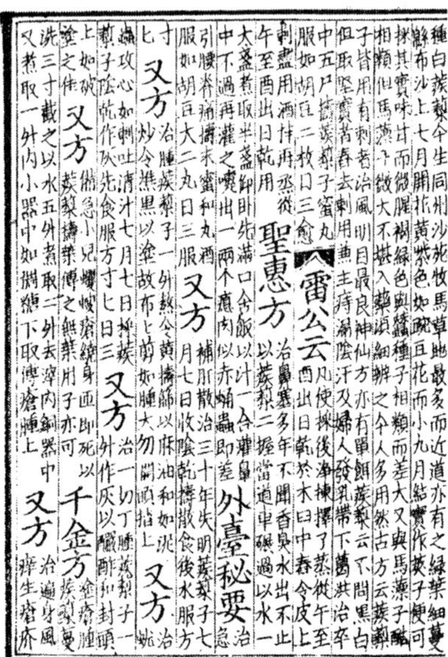

10.2 Before the use of paper, people wrote on tortoise shell bones, and bamboo.

time-consuming and laborious job until 1041–48 CE, when Bi Sheng in the Northern Song Dynasty produced a kind of movable type for printing books. Soon after, using moveable type for printing spread to countries around the world, making a great contribution to the promotion of cultural exchange and the development of human civilization.

The **compass** is a directional instrument based on the principle of magnets and is another great invention of the Chinese people. During the Warring States Period, the Chinese made use of the magnetic feature that is always pointing to the north to make a directional instrument, called a *sinan*, before the invention of the South Pointing Chariot and the Guide Fish came into being. Until the Northern Song Dynasty, people built compasses

10.3 Sinan's Spoon.

from artificial magnets, which could be used in navigation. *Sinan* is a natural magnet made through the human-made process of striking jade, creating *Sinan's* Spoon.

The Guide Fish was made by cutting thin iron into magnetized fish-shaped leaves, which could show direction. This was of great help to marching troops. The compass was more practical and convenient than the South Pointing Chariot or the Guide Fish. The invention of the compass and its application provided the conditions necessary for Navigator Zheng He in the Ming Dynasty to complete his voyage to East Africa and other places, and for Columbus' journey to North America, Magellan's global navigation, and modern sailing in general.

Gunpowder is also a kind of explosive, a material that can produce fireworks and firecrackers. The use of firearms in China began about five centuries earlier than it did in Europe. The Europeans did not know of the effect of black powder until the thirteenth century. Printing has promoted the spread and development of culture and become an outstanding contribution to the Chinese nation and world civilization. Printing later spread to North Korea, Japan, Egypt, and Europe. It was not until the fifteenth century that printing appeared in Europe. Its development took a dramatic step forward thanks to the technological advances made by Johan Gutenberg, a goldsmith working in Mainz, Germany, in the middle of the 15th century. This was four hundred years after Bi Sheng's invention of the printing press.

Papermaking spread to Arabia in the eighth century, which helped change medieval European culture, speeding up the dissemination of cultural knowledge.

Chinese Culture through Language

四大发明

Pinyin: Sì dà fā míng

Literally: Four great inventions.

Meaning: Ancient China's four great inventions: the compass, gunpowder, paper-making, and printing.

候风地动仪

Pinyin: Hòu fēng dì dòng yí

Literally: Instrument of waiting wind and land shaking.

Meaning: Ancient seismograph invented by Zhang Heng in the Han Dynasty.

指南车

Pinyin: Zhǐ nán chē

Literally: South pointing chariot.

Meaning: An ancient Chinese vehicle with a wooden figure, always pointing to the south.

CHAPTER 11

Chinese Characters

方块字

Chinese characters have a history of three thousand years. Chinese is a unique writing system because of its method of word formation and the expressive functions of shape, tone, and meaning.

Chinese characters are one of the oldest and, because of China's large population, most widely used writing systems in the word. How did Chinese characters come into being? A well-known legend says that in the period of the Yellow Emperor, more than 4,500 years ago, a legendary figure named Cang Jie created Chinese characters. He was reputedly born with four eyes and could write and observe everything in the world immediately. In truth, it was the ancestors of the modern Chinese people who created characters through hard work and civil life.

In China, people have found rock paintings of hunting scenes that are more than ten thousand years old. The shapes of deer, dogs, and basic weaponry were found to be similar to present-day Chinese characters. Thus, rock paintings played a role in the formation of Chinese characters. The symbols on ancient pottery from four thousand to five thousand years ago are also predecessors of Chinese characters. Oracle-bone inscriptions are considered the earliest "mature" Chinese characters. There are about five hundred thousand to six hundred thousand Chinese characters in total, so it is

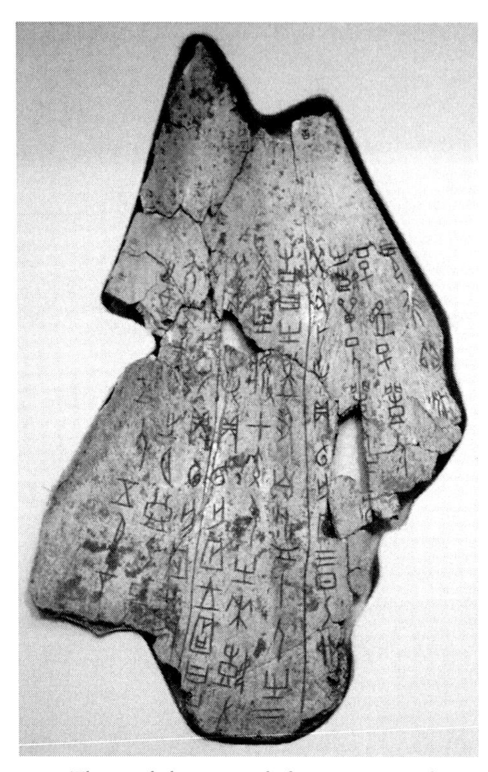

11.1 This oracle bone is made from an ox scapula dating back to the Shang Dynasty.

impossible to create all of them by drawing. Four methods for forming Chinese characters were used by Chinese people: pictographic characters, self-explanatory characters, associative compound characters, and pictophonetic characters.

Pictographic characters or pictographs represent an object in pictorial form, such as 日 (sun) and 月 (moon). Some experts and scholars believe that Chinese characters were developed from pictographs.[4] But some words could not be represented by pictographs, so words were created such as 刃 (edge), which has a dot added to the side of the word 刀 (knife). This kind of word formation is called a self-explanatory character.

The composition of two or more words belongs to associative compound characters. For example, the character 林 (forest) is composed of two trees. The character 灾 shows us a house that is on fire (the character for fire has the roof component—also known as a radical—above it), which means disaster. Because there are some limitations to this type of word formation, Chinese people tried to create new characters by combining the meanings of one character with the pronunciation of another. For instance, 忠 (loyalty), is a combination of 心 (heart) and the pronunciation of 中 (middle). This is known as the pictophonetic method for the formation of Chinese characters. These characters account for over eighty percent of all Chinese characters.

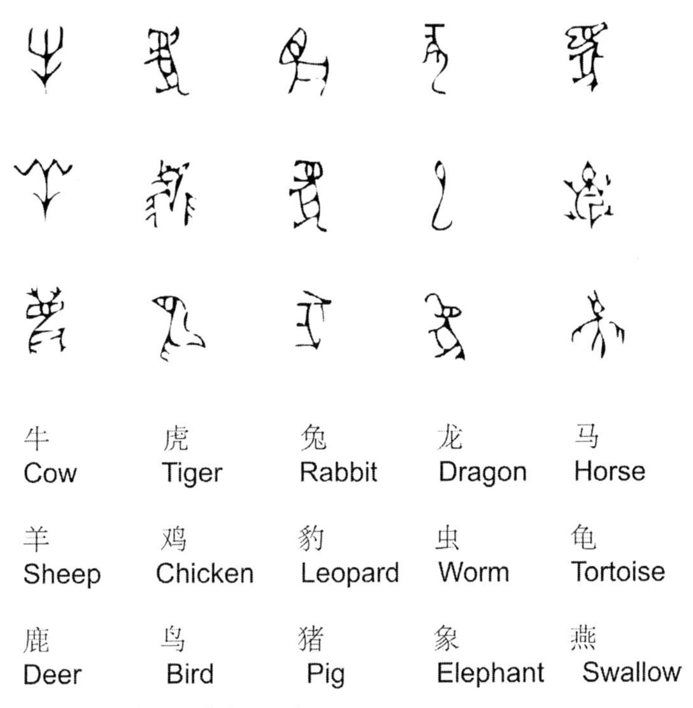

11.2 The Evolution of Chinese characters.

A unique part of Chinese characters is that almost every character reflects the thinking of ancient culture. For example, the character 安 (peace) shows that peace can be found when there is a woman in a house; the top component means roof, and the character under it is a woman. The ancient Han people believed that a quiet and comfortable family life mainly depended on women. The character 鲜 means delicious. It is a combination of the ancient Chinese regard for 鱼 (fish) and 羊 (mutton) as the most delicious foods.

Chinese characters can also reflect customs of ancient Chinese society. For example, the character 婚 means marriage, because it combines characters for woman and dusk into one. In ancient times women (女) got married at dusk (黄昏). In some places it was a traditional wedding custom for a man to marry a woman by force (also known as bride-knapping or grabbing marriage), and dusk was a convenient time for him to do so. Marriage customs like this can still be seen among some ethnic minorities now.

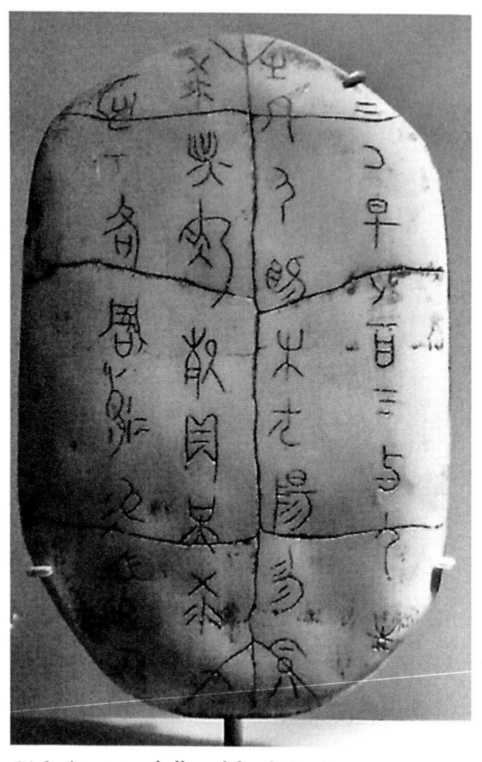

11.3 A tortoise shell used for divination.

Some Chinese characters share components and when coupled with another component create characters with similar meanings. For example, most fish have the character 鱼 (fish), coupled with another character or component. Examples include cod (鳕鱼), plaice (鲽鱼), mackerel (鲭鱼), and herring (鲱鱼); nearly all liquids have 氵 (the water radical), such as river (江), sea (海), juice (汁), and soup (汤).

During the three thousand years from Shang Dynasty to the present, the writing forms of Chinese characters have experienced many changes from the oracle-bone and bronze inscriptions, big and small seal scripts, to the official script and the regular script. However, the greatest change is the simplification of Chinese characters, which has allowed widespread development and use. Its simplicity has encouraged increased rates of literacy.

Although English and Chinese have two completely different writing styles, in word formation, a large number of affixes are used in both languages. For example, Chinese people add 电 (electric) to most objects run by electricity, such as 电视 (television), 电冰箱 (refrigerator), and 电扇 (electric fan). In English, many objects that run on electricity use electric-, such as electrical, electrician, electronic, and electrify.

Chinese Culture through Language

与君一席话，胜读十年书

Pinyin: Yǔ jūn yī xí huà, shèng dú shí nián shū

Literally: A single conversation with you is better than ten years of study.

Meaning: Talking with a wise person is better than just reading books.

开卷有益

Pinyin: Kāi juàn yǒu yì

Literally: Opening a book is beneficial (for a person).

Meaning: Reading enriches the mind.

字斟句酌

Pinyin: Zì zhēn jù zhuó

Literally: Words measuring and sentences counting.

Meaning: Choose one's words with care or weigh every word.

字正腔圆

Pinyin: Zì zhèng qiāng yuán

Literally: Clear words and correct pronunciation.

Meaning: Sing or speak with a clear and rich tone.

一字千金

Pinyin: Yī zì qiān jīn

Literally: One word is worth a thousand pieces of gold.

Meaning: This is often used to describe an excellent literary work or calligraphy.

字帖

Pinyin: Zì tiě

Literally: Chinese character book.

Meaning: Book for calligraphy; book containing models of handwriting for others to copy. They are mostly composed of stone inscription rubbings and woodcut printing.

CHAPTER 12

The Four Treasures of the Chinese Study

文房四宝

The four treasures of the Chinese study are the formal names for the writing brush, ink stick, paper, and ink slab, the essential tools of Chinese calligraphy and painting. The four treasures are varied and have a long history. The most famous are the hu bi (hu brush) made in Huzhou, Zhejiang Province, hui muo (hui ink stick) produced in She County of Anhui, xuan zhi (rice paper) made in Xuan City, Anhui, and duan yan (duan ink slab) produced in Duanzhou, Guangdong.

The writing brush was the first of the four treasures of the Chinese study to be invented. The earliest brush was unearthed from the grave of Chu, Warring States Period, in Xinyang, Henan Province, in 1957.[5] The penholder was made of bamboo and the brush was tied to the holder with string. When people first started writing they made writing brushes, called *jian hao*, with the hair of deer and sheep. The Tang Dynasty saw the peak of Chinese calligraphy and art, and better materials were selected to make writing brushes. After the Yuan Dynasty, brushes were mainly made of wool. People also used other materials, such as the hair of deer, raccoons, dogs, and tigers, wolf tails, and the feathers of geese and chickens. Some brush holders were even considered artworks, for they were made of ivory, keratin fiber, horn, jade, colored glaze, porcelain, enamel, and rosewood.

12.1 Chinese paint brushes come in many different shapes and sizes.

The *hu bi* was produced in Huzhou, Zhejiang Province. Brushes from this region are famous for their quality of material and exquisite artisanry, making them excellent for painting and calligraphy. It is said that the general of the Qin Dynasty, Meng Tian, once made a writing brush with wool in Shanlian Village, Huzhou. He has been credited with the invention of the writing brush, and the local people built a temple in his name to commemorate him.

Chinese ink originated in very early times as well. There is no doubt that the black color of ancient pottery and the inscriptions on the bones of tortoise from the Shang Dynasty were painted or written with crude ink. The early ink stick could not be ground into ink without the help of a grinder. As a result of the popularization of painting and calligraphy in the Tang Dynasty, the ink manufacturing industry underwent further development. Some masters of painting and calligraphy made ink sticks themselves and became masters Many scholars loved to collect and enjoy good ink sticks. Now a lot of famous ink sticks made in ancient times are considered works of art.

The best ink sticks, *hui mo*, were made in Huizhou, Anhui Province. They are famous for their light weight, pleasant smell, and clear and deep black

12.2 A Chinese calligraphy set consists of brushes, two kinds of ink, an ink slab, and a chop.

when ground. The emperor at the time, Li Yu (711–762), appreciated *hui mo* very much and granted his own family name, Li, to the producer, which was a great honor. Since then the Li ink stick has become very famous. There was an popular saying at the time: "It is easy to get a thousand pieces of gold, but it is difficult to buy a Li ink stick."

Paper is one of China's greatest inventions. Before its invention, Chinese characters were engraved on bones and tortoise shell, later on bronze ware and stone, and then on bamboo bark and thin silk. In the middle period of the Eastern Han Dynasty, a person by the name of Chai Lun invented a type of paper and was praised by the emperor, who ordered its promotion throughout the country. Later, Chinese people made different kinds of paper according to different specificities.

The four most common types of paper are described here.

Hard yellow paper can be very long and is able to maintain its form, so in the past it was used to copy Buddhist Scriptures, many of which have been perfectly preserved. *Jin hua* paper (golden flower paper) was mainly used for decoration. In the process of making such paper, gold was ground into powder, which was sprinkled on colored paper or sticky paper. *Mao bian* paper was made from bamboo, which was most appropriate for printing a large number of books at a time. Rice paper is the most distinctly Chinese. It was offered as an article of tribute to emperors or courts in many dynasties. Rice paper is mainly used for traditional Chinese painting and calligraphy: It is soft but tensile and resistant to insect bites, absorbs ink evenly, and is good for long-term preservation. The style of Chinese painting and calligraphy is closely related to the properties of rice paper.

The ink slab is a uniquely Chinese writing tool, used to grind ink sticks for writing. Ink slabs have many kinds of shapes and can be made of different materials. After the Han Dynasty, most ink slabs were round, and pottery ink slabs were very popular. During the Sui and Tang Dynasties, triangular and tortoise-shaped ink slabs were mainly used. The most popular were dustpan-shaped ink slabs. Since the Song Dynasty, most ink slabs have been made of stone. The four famous ink slab varieties in China are *duan yan* (*duan* ink slab), *she yan* (*she* ink slab), *taohe yan* (*taohe* ink slab), and *dengni yan* (*dengni* ink slab). *Duan yan* is considered the best because of its ability to keep the ink

12.3 A traditional duan yan ink slab.

moist, its durability, and its beauty; it is made of colorful materials and does no damage to the brush.

Although the production of Chinese brushes and ink is quite different from that of the West, the uses are similar. For example, the goose quills that Westerners used long ago had to be dipped in ink before use just like Chinese brushes. It is said that the fountain pens we use now were invented by an American entrepreneur named Lewis Waterman more than one hundred years ago. The quills he made could not deposit ink cleanly and often dirtied the writing paper. He therefore had trouble in his business, so he decided to improve the pen. First, he added a reservoir to store ink and then designed a special writing point. Since then, pens have replaced quills, which were used by Europeans and North Americans for many years.

Chinese Culture through Language

纸包不住火

Pinyin: Zhǐ bāo bù zhù huǒ

Literally: Paper can't wrap up a fire.

Meaning: There is no way to conceal the truth.

罄竹难书

Pinyin: Qìng zhú nán shū

Literally: Not enough bamboo strips to write on.

Meaning: Too many misdeeds to be recorded.

投笔从戎

Pinyin: Tóu bǐ cóng róng

Literally: Lay aside the writing brush and join the army.

Meaning: To give up scholarship and become a soldier.

CHAPTER 13

Lute, Chess, Calligraphy, and Painting

琴棋书画

The lute, chess, calligraphy, and painting are described as the four major classical artistic forms or arts in China. There are similarities among them; all should be used in a way that demonstrates authenticity and balance, all in the pursuit of conveying depth, spirit, and loftiness. They are symbols of elegance, the soul of Chinese culture.

The lute, chess, calligraphy, and painting, are known as the four arts, all of which the ancient Chinese literati admired, respected, and mastered.[6] More than two thousand years ago, Confucius advocated people master six arts, two of which are music and calligraphy. In fact, the four mentioned here are the refinements of the six artistic forms. In ancient China, all the literati loved them, and they became a symbol of the literate and educated.

The lute (*qin*), commonly known as *guqin*, is one of the oldest musical instruments in China. It is one of the oldest known stringed instruments and is said to be the father of Chinese music. According to literary records, the *qin* has more than three thousand years of history. It is 130 centimeters long, 20 centimeters wide, and about 5 centimeters thick. In total there are seven strings, and on the fret panel there are thirteen small round emblems used to mark the location of pitches. The *qin* has a wide range, up to four octaves,

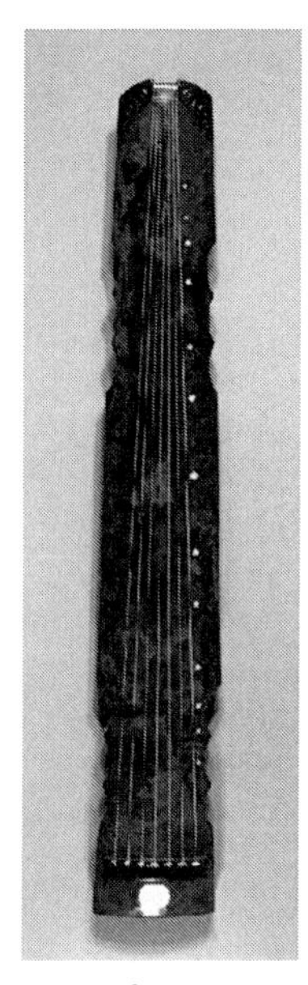

13.1 This qin is part of the zither family.

meaning one can play many variations of rich sound. The *qin* is representative of the ancient scholars.

Chess (***qi***), in ancient China, mainly refers to *xiangqi* and *weiqi* (types of Chinese chess). *Weiqi*, sometimes called Go in English, originated from China and has more than two thousand years of history. The action of playing *weiqi* is called *duiyi*, or *yi*; both sides are always attacking and surrounding each other. The chess pieces or stones are divided into black and white, symbolizing the principle of yin and yang. There are 181 black stones and 180 white, for a total of 361 stones, exactly the number of points as are on the chessboard. Chess players have to put the pieces on the intersections of the lines, and they cannot be changed after being set down. The player with the black stone has the first move, both sides then taking turns; only one piece may be put down at a time. The winner is determined

13.2 Chess is a popular past time. Many parks have stone chessboards where people gather to play.

by the number of colored pieces on the intersections. Black must occupy more than 184 intersections, and white over 178. The rules of Chinese chess are simple, but its simplicity and an individual's wit depict the bright essence of Chinese culture within Chinese chess.

Chinese chess conveys a profundity and mystery, so many rulers and even common people became interested in playing it. It has also inspired many legendary and charming stories, poetry, and even books on military strategies. Governing policies have also often been influenced by Chinese chess. It is a hallmark of Chinese civilization and history.

Calligraphy (***shu***), specifically Chinese calligraphy, is a unique Chinese artistic skill. It is sometimes accompanied by drawing, which is known as calligraphy and painting. It is a kind of art that takes the Chinese character as the object, and a brush as a tool. After years of development, Chinese calligraphy has become a marvelous art form, with a unique style that changes constantly continues to evolve.

The history of calligraphy may be roughly divided into five categories: the development of the seal script (big seal and small seal scripts), official script, regular script, running script, and cursive script.

Seal script: The big seal was the precursor to the small seal script. The big seal script appeared in the Western Zhou Dynasty; the strokes are powerful and dignified, the structure is square-shaped, defining its classic style. After Emperor Qin Shi Huang unified China, the typeface was simplified and unified, and people called it *xiaozhuan* (small seal script). The typeface was long and had neat circle-strokes as well as an elegant form.

Official script appeared as early as the Eastern Han Dynasty, and it is the basis of today's writing. It caused the revolution that changed the shape of Chinese writing from round to square, and joined lines to individual strokes, making writing easier. This style was popular among the lower court officials; therefore, it was called official script. The appearance of official script was a turning point in the evolution of Chinese calligraphy and laid a solid foundation for regular script.

Regular script is a type of calligraphy that has gradually evolved from official script. It streamlined its shape from the original flat size; this script became tidy and formal and eventually got the name *kaishu*. It can be used as a sample type to press a book and is still in use today. Regular script was prevalent in the Six Dynasties and reached its peak in the Tang Dynasty. This kind of script is still the most important for beginning calligraphers.

Running script is used to write regular script quickly. When written similarly to regular script, it is called *xingkai*; when similar to cursive script, it is called *xingcao*.

Cursive cannot be regarded as arbitrary scribble, for it is based on its own principles. Its greatest feature is its artistic value, which is superior to its practical value. Generally, it can be classified into two types: *zhangcao* and *jincao*. *Zhangcao* is a kind of official sketch; every character is independent but not written continuously, whereas *jincao* is a fast, regular script with up and down strokes between characters.

Chinese calligraphy uses variations in lines to demonstrate various styles and elegance. The Chinese character, the brush pen, and cultural thought have created a unique artistic world of calligraphy.

13.3 Calligraphy is taught in schools, and is a means for both artistic expression and the memorization of characters.

Painting (***hua***), national or native painting, has distinctive ethnic features and a long history. It is a valuable cultural legacy for the world. In the Warring States Period more than two thousand years ago during the Chu Dynasty, two famous paintings on silk relied on vivid patterns, a simple style of drawing, and smooth lines to express a perfect wonderland: These paintings established the art of Chinese painting and shaped the national style.

Traditional painting is done with a brush dipped in black or colored ink on fine rice paper or silk. Paintings can be divided into three types depending on the content: figure painting, landscape painting, and flower and bird painting. Figure painting reached maturity in the Warring States Period and was at its peak in the Tang Dynasty. Landscape painting is based on showing natural beauty. Flower and bird painting focuses on flowers, birds, bamboo, fish, and insects as its imagery. There are mainly two techniques in Chinese painting, both of which are meticulous; *gong bi* is often is referred to as court-style painting and is very detailed, and *shui mo* is loosely termed water color or brush painting. The Chinese character *mo* means ink, and *shui* means water. This style is also referred to as *xieyi* or freehand style. The characteristics of court-style painting are lifelike depictions in great detail with many strokes. Freehand is a kind of exaggerated painting; the works are simple but vivid. Chinese painting focuses on balance and harmony to create unique poetic scenes and is a representative style of the East, different from Western art styles.

Ability to play the lute, play chess, write calligraphy, and paint are the four main accomplishments required of Chinese scholars. These particular skills are unique among the arts of the world. Chinese calligraphy is based on Chinese characters, which dates to the origins of recorded Chinese history, in essence ever since writing has existed. Chinese calligraphy is said to be an expression of a practitioner's poetic nature. It has evolved for thousands of years and differs from Western calligraphic script in the sense that it is done with a brush instead of metal implements or a quill. Calligraphy, as well as the other three arts, is the means by which a scholar could compose and communicate his or her thoughts and thus be immortalized. It was the scholar's means of creating expressive poetry and sharing his or her own learnedness. The four arts are treasures of traditional Chinese culture, a way to cultivate people's minds and to promote mental health.

Chinese Culture through Language

白纸黑字

Pinyin: Bái zhǐ hēi zì

Literally: White paper and black words.

Meaning: Written in black and white.

举棋不定

Pinyin: Jǔ qí bù dìng

Literally: Holding a chess piece and hesitating.

Meaning: hesitate about what move to make.

文如其人

Pinyin: Wén rú qí rén

Literally: The writing is like the author.

Meaning: The style is like the man, the writing reflects the author.

扬州八怪

Pinyin: Yáng zhōu bā guài

Literally: Eight Eccentrics of Yangzhou.

Meaning: Eight Eccentrics of Yangzhou who made a great contribution to Chinese arts (Jin Nong, Wang Shishen, Huang Shen, Li Xian, Zheng Xie, Li Fangying, Gao Xiang and Luo Pin). They were all living and working in Yangzhou during the reign of Emperor Qian Long of the Qing Dynasty. They were described as "eccentric" for their unconvential painting, writings, temperament, and behavior, and created new styles of painting which made a great contribution to painting and the arts.

柳体

Pinyin: Liǔ tǐ

Liu style (a calligraphic style created by Liu Gongquan of the Tang Dynasty).

欧体

Pinyin: Ōu tǐ

Ouyang style (a style of calligraphy represented by Ouyang Xun of the Tang Dynasty).

CHAPTER 14

Chinese Medicine

中医

China is one of the earliest countries to develop medicine. The early production of a large number of classical Chinese medical books written by famous doctors was a rarity in early times. For thousands of years, traditional Chinese medicine has been used to cure a lot of diseases and alleviate suffering for Chinese people and has made a tremendous contribution to the growth and prosperity of China.

The formation of traditional Chinese medicine (TCM) can be traced to the primitive period (before the 21st century, BC).[7] One can read in ancient literature that Shen Nong, the father of Chinese agriculture and medicine, tasted hundreds of herbs in order to find new medical treatments. Early peoples gradually invented medical treatments, implements, and medicines.

Examples include the stone needle, moxabustion therapy, and herbal soup. People have used the sharpened sides of stones as acupuncture needles to cut carbuncles or to stimulate certain parts of the body to eliminate pain. Moxibustion is a process of alleviating pain or suffering with heated stones and sand. People also developed a method of cooking herbs to treat diseases. Herbal soup is a common remedy; herbs are boiled in water or alcohol, and then the decoction is consumed. It is the earliest and most widely used method of medical treatment in China.

14.1 Many Chinese herbal shops display their wide range of herbs in glass jars.

There are four diagnostic methods in Chinese medicine: observation, auscultation or smelling, interrogation, and pulse taking and palpation. Observation involves observing the patient's face, body, tongue, and urine; auscultation is listening to the patient's voice, breathing, moaning, and coughing. Smelling refers to sniffing the patient's breath and urine; interrogation is questioning the patient about his or her background, such as the duration of the disease, symptoms, and living habits. Pulse taking and palpation mean reading the pulse, checking the skin, the chest, and the abdomen. According to the theories of Chinese medicine, the four methods of diagnosis should be used together for the correct diagnosis and proper treatment.

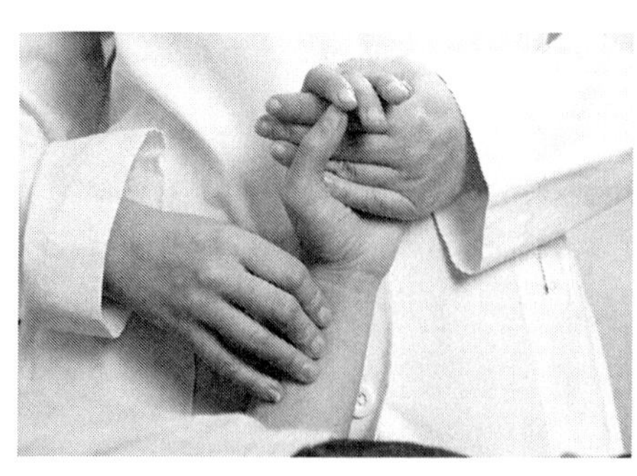

14.2 The quality of a patient's pulse helps the doctor diagnose a patient's illness.

The earliest Chinese medical book is *The Yellow Emperor's Canon of Internal Medicine*, required reading in Chinese medicine for thousands of years. The book discusses the details of the theoretical knowledge of yin and yang, and the five element principles of Chinese medicine: *zangfu* (the pairing of yin and yang organs), *jingluo* (channels of energy distribution in the body), causes of disease, diagnosis, treatments, and so on.

The first book of pharmacology was published in the later Han Dynasty. It summarizes early medical experience and documents 365 kinds of herbs and their places of production, character, collection, and the major functions of every medicine in great detail. The book had a far-reaching influence and offered great guidance to the development of Chinese medicine.

Four great Chinese works on medicine are *The Yellow Emperor's Canon of Internal Medicine*, *Shen Nong's Canon of Herbs*, *Shang Han Lun*, and *Jin Gui Yao Lue*. The last two were taken from *Shang Han Za Zheng Lun*, which was written by a famous medical scientist, Zhang Zhongjing. The book analyzes medical cases using the four methods of diagnosis from TCM and contains more than three hundred prescriptions. These prescriptions are the predecessors of modern prescriptions and are highly valuable.

There are many famous doctors in Chinese history, such as Bian Que, Hua Tuo, Sun Simiao, and Li Shizhenn. Bian Que was born during the Warring States Period. He was able to cure many diseases and was the founder of pulse reading and observation. He was very good at these skills. According to many stories, he cured sickness and saved patients. Once, as he was passing through the State of Guo, Bian Que heard the Prince of Guo had died. Bian Que entered the palace and observed the body of the prince carefully. He believed that the prince had not died, so he used acupuncture to treat him. The prince then regained consciousness. From then on, people spoke of him as a highly skilled doctor.

Hua Tuo was a famous doctor in the Eastern Han Dynasty. He was good at surgery and invented anesthetics. Patients who took his *mafo an*, an anesthetic, could not feel any pain during operation.

Sun Simiao was a famous medical expert of the Tang Dynasty. He specialized in the treatment of various diseases and cured many patients on the brink of death. On one occasion, he saved a pregnant woman who had already been placed in a coffin, and helped her give birth. He also wrote two medical books, *Qianjin Prescriptions* and *Qianjin Yifang*, which includes more than six thousand prescriptions. He is called the king of Chinese medicine.

Li Shizhen was a well-known scientist of Chinese medicine during the Ming Dynasty. He wrote the *Compendium of Materia Medica*, a lengthy Chinese medicine masterpiece. He spent thirty years walking around the mountains and rivers of China, risking his life to collect and taste a variety of herbs. The book records 1,892 kinds of herbs and more than 1,000 prescriptions. It is an important work for the development of medicine and pharmacology throughout the world and has been translated into many languages.

More and more people are paying close attention to the medical effects of TCM. Chinese medicine has made a great contribution to the progress and development of world medicine.

Traditional Chinese medicine and Western medicine are very different in their methods of examination and treatment. The doctors of TCM give diagnoses by observation, auscultation and smelling, and pulse taking and palpation. They often look at the illness from a macroscopic angle and focus on a treatment for the whole body. Doctors of Western medicine focus on body temperature and laboratory results, and treat the patients from a microcosmic point of view. Natural herbs and acupuncture are used in TCM, whereas Western medical practitioners prefer to use chemical and synthetic drugs and surgery. At present, it is common for people seaking medical treatment to choose Western medicine to treat an illness and Chinese medicine to recover from it.

14.3 Herbs commonly used in Chinese medicine.

Chinese Culture through Language

扶正祛邪

Pinyin: Fú zhèng qū xié

Literally: Strengthen the good and repel the bad.

Meaning: To strengthen the body resistence against disease and dispel pathogenic factors.

灵丹妙药

Pinyin: Líng dān miào yào

Literally: Miraculous pellets, magic drug.

Meaning: A magic medicine or cure-all: a remedy for all problems.

望闻问切

Pinyin: Wàng wén wèn qiè

Literally: Look, listen, ask, and feel the pulse.

Meaning: These are the four methods of diagnosis used in traditional Chinese medicine.

同病相怜

Pinyin: Tóng bìng xiāng lián

Literally: Same illness, have sympathy for each other.

Meaning: People with the same situation or suffering sympathize with each other.

针灸

Pinyin: Zhēn jiǔ

Literally: Needles and moxibustion.

Meaning: Acupuncture and moxibustion.

CHAPTER 15

The Chinese Way to Stay Healthy

养生文化

China has a civilization with five thousand years of history and much great culture. Part of this culture concerns staying healthy. The main purpose is to keep people fit and living longer. It was developed through many tried and tested methods of Chinese people.

The earliest experts at staying healthy were the Chinese philosophers Confucius and Zhuangzi. Confucius popularized methods of having a healthy diet and stressed the need to foster morals. *Xiao Yao You,* written by Zhuangzi, played an important role in the fostering of moral values and staying healthy through *qigong*, the practice of aligning breath, movement, and awareness for exercise, healing, and meditation.

The best books on how to cultivate health are *The Yellow Emperor's Canon of Internal Medicine*, *Master Lu's Spring and Autumn Annals*, and *Xing Qi Yu Pei Ming*. The first book emphasizes the principles between humans and nature. The second suggests that people should keep healthy by adhering to the principles of the four seasons. *Xing Qi Yu Pei Ming* advises that *qigong* exercises and medical *qigong* should be used to stay healthy.[8] These books have had a significant effect on the development of health culture in China.

According to Chinese health experts, Chinese ways to stay healthy can be divided into two types. One is the maintenance of the body, known as "the

15.1 Tai chi is a gentle form of martial arts favored by many elderly people to maintain their health.

way to maintain a good physique." The other is the maintenance of the spirit, which is called "relaxation."

In order to keep a good figure, people must perform physical exercises. An ancient and renowned doctor, Hua Tuo, created the five-animal exercises, physical exercises to prevent disease and extend one's life. The five-animal exercises is a set of exercises imitating the movements of the tiger, deer, bear, ape, and bird. Although regular exercise can help people keep fit, it must be moderate. The appropriate amount of exercise is the key to keeping in good health, because excessive exercise may be harmful.

Another key part of keeping in good health in China is relaxation. There are many ways to rest to attain tranquility. The Daoist philosopher Laozi believed that the key is to keep quiet; most importantly one must be tranquil, not think too much, and have few desires.

15.2 Some doctors still use a mortar and pestle to prepare herbal remedies.

Chinese medicine also places emphasis on diet. Most Chinese people are firmly convinced that the "kitchen physic [medicine] is the best physic" and if the balance of food is not right, it will be harmful to one's health. The Chinese diet requires a light diet and no rich food, eating both refined and coarse grains, and consuming a consistent quantity of beans and vegetables at fixed times. One should not eat or drink too much. There are many old sayings about diet and health; for example, "Eating radishes can keep the doctor away," "If we want to keep fit, we should not eat too much," and "Many dishes make many diseases."

The Chinese diet has a close relationship with the five flavors (sour, bitter, sweet, spicy, and salty), the five internal organs (liver, heart, spleen, lungs, and kidney), and the four seasons (spring, summer, autumn, and winter). In spring, more sour- and sweet-tasting food should be eaten because sour-tasting food is good for the liver, and sweet food can nourish the spleen. In summer, bitter is the main flavor to eat because most bitter-tasting vegetables can relieve internal heat and eliminate fatigue. It is said that hot is a good flavor in autumn, and salted food is suitable for winter. In the dry autumn, people ought to eat more fruit and vegetables, such as pears, grapes, Chinese dates, lilies, and white fungus, which can nourish the lungs, the organ correlated with autumn.

The old ways to stay healthy in China are extensive and profound. If people study and practice them carefully they can preserve their health, eliminate disease, and even prolong life.

There are many similar sayings in China and in Western countries about how to stay healthy. For example, an old Chinese saying goes: "If you want to keep fit, leave [a meal] with an appetite and dress appropriately" (you must dress appropriately according to the season). There are many food and health related proverbs in other languages and cultures. Inventor and founding father of the United States Benjamin Fraklin said "Many dishes make many diseases."A French proverb says: "Greedy eaters dig their graves with their teeth." Galen of Pergamon, a Roman of Greek ancestry said "More are killed by gluttony than by the sword." Therefore, in order to keep fit, people from the West and East value a proper diet.

Chinese Culture through Language

一日一苹果，医生不找我

Pinyin: Yī rì yī ping guǒ, yī shēng bù zhǎo wǒ

Literally: An apple a day, the doctor will not find me.

Meaning: An apple a day keeps the doctor away.

病从口入

Pinyin: Bìng cóng kǒu rù

Literally: Diseases enter through the mouth.

Meaning: Sickness is caused by what we eat. This can refer to eating too much food, poor quality food, or unclean food.

心宽体胖

Pinyin: Xīn kuān tǐ pàng

Literally: Wide heart, fat body.

Meaning: To be joyous and carefree.

CHAPTER 16

Wushu

中国武术

Wushu (Chinese martial arts) has thousands of years of history, many schools of study, and is a symbol of traditional Chinese physical philosophy. In ancient times, the main purpose of wushu was to defend against the enemy in military affairs, but nowadays people practice it just to maintain good health.

The origins of *Wushu* (Chinese martial arts) go back to primitive society. In the harsh struggle for survival, some attacking and defending skills were needed. In an effort to improve physical strength and advance their skill in hunting, early peoples developed *wushu*.[9] In a slave and feudal society, people developed fighting skills quickly as wars broke out frequently. The turbulence motivated people to practice *wushu*.

Wushu is a defensive and offensive combat system that includes many set patterns such as kicking, hitting, throwing, catching, chopping, and stabbing.

During China's long history, *wushu* has been developed and perfected. There are many kinds of martial arts and different factions or groups. Two main boxing skills were formed in the Ming and Qing Dynasties, *Shaolin* and *Wudang*. Long boxing, short boxing, and *ditang* boxing belong to the *Shaolin* style and are mainly used for fighting. Only taichi boxing and eight trigram belong to *Wudang* boxing, mainly used for defending oneself and keeping in good health.

16.1 Two Shaolin monks demonstrate their weaponry mastery.

Another important aspect of *wushu* is weaponry skills. *Wushu* weapons used include broad swords, swords, spears, cudgels, axes, hammers, forks hooks, and halberds.

During the Sui and Tang Dynasties *wushu* reached a new level of prosperity. From that time, it became common practice to choose outstanding talents from martial arts competitions as military commanders. Many ancient heroes like Yue Fei were chosen from these competitions. This practice also improved the practice of martial arts in the whole country. At that time, soldiers practiced martial arts, as did scholars. It was said the most famous poets in China, Li Bai and Du Fu, also learned how to use swords when they were young.

The Shaolin Temple's rise to prominence is an important part of *wushu* history. The temple is located on Songshan in Dengfeng County, Henan Province. It is said that during the Tang Dynasty, the monks helped Li Shimin, who later became emperor, and were handsomely rewarded for their contribution. Since then, the kung fu (also known as gong fu, another form of martial arts) of Shaolin has been developed and become famous all over the world.

16.2 Men and women, Chinese and foreigners practice Wudang tai chi in the misty morning.

After the Southern Song and Northern Song Dynasties, *wushu* became popular among ordinary people and became an important skill to maintain good health and defend the homeland. As time went on, the military function of *wushu* gradually became less important, but it became a popular way to keep fit. There has been a close relationship between martial arts and methods to maintain good health, especially after the Qin and Han Dynasties. Its practical functions have enabled *wushu* to continue to flourish.

Wushu pays great attention to harmony from inside and outside and combines people's actions with their environment. People usually use something within their natural environment as a frame of reference when they practice it. The forms are often described as looking like a monkey, standing like a tree, sitting like a bell, being as light as a leaf and as quick as the wind.

Nowadays, many physical education colleges and universities have *wushu* majors. There are a lot of *wushu* enthusiasts in China and abroad. It was part of the 2010 Asian Games and continues to gain popularity.

It is well known that Chinese and Western arts of attack and defense techniques are mixing together gradually. For example, the hybrid combat sport of kickboxing, has become popular in the United States. It is a modern style of combat, which is formed on the basis of the mutual exchange and interflow of both Western and Asian martial arts. Jeet Kune Do, created by Bruce Lee, is another example of the integration of Chinese and Western martial arts.

Chinese Culture through Language

害人之心不可有，防人之心不可无

Pinyin: Hài rén zhī xīn bù kě yǒu, fáng rén zhī xīn bù kě wú

Literally: Don't harbor a heart to hurt others; but don't neglect to have a heart to guard against others.

Meaning: Never harbor a mind to harm others, but always be on guard against being harmed by others.

以其人之道还治其人之身

Pinyin: Yǐ qí rén zhī dào huán zhì qí rén zhī shēn

Literally: Treat that person using his/her way of treating others.

Meaning: Give someone a taste of their own medicine.

路见不平，拔刀相助

Pinyin: Lù jiàn bù píng, bá dāo xiāng zhù

Literally: See injustice on the road and draw one's sword to offer help.

Meaning: Help others in need.

知彼知己，百战不殆

Pinyin: Zhī bǐ zhī jǐ, bǎi zhàn bù dài

Literally: Know the enemy and yourself, and you will win a hundred battles without defeat.

Meaning: If you have knowledge of your enemy and yourself, you will always find success.

强中自有强中手

Pinyin: Qiáng zhōng zì yǒu qiáng zhōng shǒu

Literally: There is always a stronger one among strong hands.

Meaning: There is always someone stronger.

化干戈为玉帛

Pinyin: Huà gān gē wéi yù bó

Literally: Turn weapons of war into jade and silk fabrics.

Meaning: beat swords into plowshares.

十八般武艺

Pinyin: Shí bā bān wǔ yì

Literally: eighteen martial arts of ancient Chinese weapons.

Meaning: To have many various skills.

文武双全

Pinyin: Wénwǔ shuāngquán

Literally: Good at both arts and martial arts.

Meaning: Be well versed in both polite letters and martial arts; adept with both pen and sword.

CHAPTER 17

Traditional Chinese Weapons
中国古代武器

Traditional Chinese weapons refer to offensive weapons and protective equipment used by the Chinese military and civilians from the late primitive society (before the Xia dynasty) to the late Qing Dynasty. During ancient times there were many fights among clans and tribes, so tools with sharp points or cutting edges were used as weapons. By the late Neolithic period (5,000–2,000 BCE) people had mastered the skills of making polished stone tools and started to make offensive weapons, many of which were long-range weapons and combat weapons; they also made body armor.

Hundreds of kinds of ancient Chinese weapons were invented following technical developments and discoveries in materials and processes. There are three types: short weapons, long weapons, and hidden weapons.

Generally speaking, the length of the short weapon was not taller than an ordinary person's eyebrows and was light enough to be used with one hand. Knives and swords are the most common short weapons. There are two kinds of swords in China, the *wen* sword, which has a tassel at the handle, and the *wu* sword, without a tassel. The mace, hook, and ax are types of short weapons.

The most common long weapons are spears, machetes, and staffs. The spear is the king of them all. A lot of heroes in China used spears, such as the famous generals Yue Fei and Yang Zaixing in the Song Dynasty. The broadsword has

17.1 A variety of Chinese long weapons.

17.2 The guan dao, a pole weapon, was invented in the 3rd century.

a long-handled blade, usually three meters long and 7.5 kilograms (16.5 pounds) in the Tang Dynasty. The weapon used by the famous general Guan Yu in the story *Romance of the Three Kingdoms* is a type of broadsword known as the crescent moon-shaped sword.

There are many types of sticks such as long sticks, *qimei* sticks, and the three-section cudgel. Many monks of the Shaolin Temple use sticks as weapons. It is said that once a group of bandits invaded the Shaolin Temple and defeated many monks. A monk with a lit stick jumped out of the kitchen and trounced the bandits. Since then, the Shaolin Temple monks have used sticks as weapons. Halberds, forks and shovels, and palladiums, which were farm tools, also belong to the category of long weapons.

Hidden weapons make it easy to carry out an attack in secret. They are small, portable, quick, and very powerful when used at close range. There are four types: hand-thrown, cable hitting (a Chinese technique), machine-firing, and spraying (the spraying of poison or other noxious substance).

Hand-thrown hidden weapons, among which are javelins, darts, money darts, flying knives, stones, needles, and flying rings, are used most extensively. The ancient people who were good at kung fu often used pocket coins as hidden weapons. The flying ring was made of iron and had a diameter of fifteen centimeters. People adept in *wushu* could throw two of them at once. In the classical Chinese novel *Journey to the West*, Nezha, the third prince, used a flying ring as a weapon and often defeated his enemies with it.

The cable-hitting concealed weapon has a long rope and can be thrown and retracted at will. The most common are the meteor hammer, rope barbell, and the soft whip. The machine-firing hidden weapon is the arrow in the sleeve. People tie the arrow case to their arms, covering it with the

17.3 Young Shaolin monks perform with three types of short swords.

sleeve. When the spring lock is released, the arrow can be fired; some can fire continuously.

In China, most people who practice martial arts can use a variety of weapons. Some are adept in the eighteen kinds of martial arts, which means that they have skills in wielding the eighteen kinds of ancient Chinese weapons. Different dynasties had different views about the eighteen kinds of weapons, and each dynasty favored certain weapons. Therefore, these eighteen generally refer to all ancient Chinese weapons.

In the East and the West, a lot of weapons evolved from the tools used in daily life. For instance, both the European crossbow and the bow from China evolved from the bow and arrow that were used for hunting. When peasants revolted, the most available weapons were the farm tools around them. The scyberd (half scythe and half halberd) evolved from the long-handled sickle that was used by the Europeans during the wheat harvest. The ax was originally a common tool in Europe; it was used as a weapon by foot soldiers.

Chinese Culture through Language

明枪易躲，暗箭难防

Pinyin: Míng qiāng yì duǒ, ànjiàn nán fáng

Literally: Revealed weapons are easy to avoid, but it is hard to stop hidden rockets.

Meaning: False friends are worse than open enemies.

百步穿杨

Pinyin: Bǎibùchuānyáng

Literally: Shooting a willow leaf; to shoot an arrow through a willow leaf one hundred paces away.

Meaning: It is widely used to describe an expert marksman regardless of whether he uses an arrow, stone, knife, or gun.

班门弄斧

Pinyin: Bānménnòngfǔ

Literally: A carpenter shows his proficiency with an axe before Lu Ban—a master carpenter.

Meaning: To offer to teach fish how to swim. This idiom is used either to ridicule someone who displays his slight skill before an expert or to express one's modesty when demonstrating a skill in front of colleagues.

笑里藏刀

Pinyin: Xiàolǐcángdāo

Literally: A knife hidden under you smile.

Meaning: A dagger can be concealed in a smile.

一箭双雕

Pinyin: Yī jiàn shuāng diāo

Literally: Catch two pigeons with one bean.

Meaning: Kill two birds with one stone.

鸟尽弓藏

Pinyin: Niǎo jìn gōng cáng

Literally: Get back the bow when there are no more birds.

Meaning: To cast someone aside when they are no longer needed.

自相矛盾

Pinyin: Zì xiāng máodùn

Literally: Attacking one's shield with his own spear.

Meaning: Paradox, to self-contradict.

CHAPTER 18

Chinese Names

中国姓名

Chinese naming is of paramount importance in Chinese life. From early times it has played a big role in various fields such as politics, culture, and social activities.

Naming is an important component of traditional Chinese culture. In the development of Chinese society, it has various functions and has played different roles in different times. The development process of Chinese names not only reflects the vicissitudes of five thousand years of Chinese history but also the origins of Chinese culture.

In ancient China, naming was very complicated, and the process took one's clan and dwelling place into consideration. Surnames, *xing* (姓) represented a common blood relationship in one's ancestry, and *shi* (氏, sub-surnames, no longer differentiated in Chinese names) were derived from one's clan area. The purpose of adopting *xing* was to determine if it was appropriate for one person to marry another. If a boy and a girl had the same *shi* but their *xing* was different, they could marry each other. However, if the opposite was true, they were not allowed to get married.[10] This shows that Chinese people knew that marriage between close relatives might produce unhealthy and less intelligent offspring. During the Han Dynasty, *xing* and *shi* became integrated as *xing*, a single family name. Chinese surnames nowadays are mostly passed down from generation to generation and have been in use for thousands of years.

There are thirteen kinds of surnames. First, *xing* and *shi* were the same in matriarchal society; people took their surnames from their mothers' names, so many of the old names have the character for woman, *nu* 女, as the left part of the surname character (this character becomes a component or radical to form the surname). Examples include Yao (姚), Ji (姬); even the character for suname, *xing* (姓), is formed with woman (女) and birth (生). In the case of the surname Jiang (姜), the character for woman is the lower component. Many surnames are derived from the names of the ancient states, such as Song, Qi, Lu, and Zheng; titles like Sima, Sikong; landforms and physiognomy near one's birthplace; natural scenery like *chi* (lake) and *liu* (willow); names of professions such as *tao* (potter); or the names of one's ancestors. An example is the Huangdi Emperor, whose name, Xuanyuan Shi, became a family name. The names of totem animals such as dragons, oxen, horses, or plants like poplars and plums, also became surnames.

Chinese people generally have one family name and two or more given names. A surname with a one-character name is a single surname, and surnames with two or more characters are compound surnames. Two words or more are combined to make surnames. There is no exact number of surnames recorded. *Baijia Xing, One Hundred Surnames*, a book complied in the Northern Song Dynasty, lists 472 surnames, beginning with Zhao, Qian, Sun, and Li, and 60 compound surnames. Statistics from the book *Chinese Surnames and Their Anecdotal Stories* show there are 6,362 surnames in literature. The most common family names are Zhang, Wang, Li, Zhao, and Liu. Zhuge, Ouyang, Sitou, and Sima are the most common compound surnames.

Chinese names are written with the family name followed by the given name. In ancient China, names were more complicated than they are today. People had *zi* (formal or adult name) and *hao* (pseudonym) in addition to *xing* and *ming*. Ancient Chinese people's *ming* (given name) and *zi* were often related, and they complemented each other. All Chinese traditionally received a milk name or nickname at birth and a formal name when entering school. In ancient China, when boys were twenty and girls were fifteen, they would be given formal names. The literati or people with titles had *hao*. *Hao* was a person's complimentary address or honorific, similar to today's pseudonym. The ancient Chinese scholars took their interests or the places in which they lived as their *hao*. For example, Li Bai, a poet in the Tang Dynasty, lived in a village called Qinglian as a child, so he used Qinglian Jushi as his *hao*. In those times people called themselves by their given names, whereas one called others by their *zi* or *hao* to show their respect.

Chinese names have special meanings and show the parents' wishes and expectations. Girls' names reveal that their parents want them to be nice, whereas boys should be strong and brave. Some names represent a time, place, or natural phenomena from the time when they were born such as *jing* (京, Beijing), *chen* (晨, morning), *dong* (冬, winter), or *xue* (雪, snow). Others wish for virtues such as *zhong* (忠, loyal) and *li* (礼, polite). Some names represent health, longevity, and happiness, *jian* (健, healthy), *shou* (寿, longevity), *song* (松, pine tree), and *fu* (福, happiness). Masculine names often show bravery, such as *hu* (虎, tiger), *long* (龙, dragon), *gang qiang* (刚强, strong and great). Feminine names are often *feng* (风, wind), *hua* (花, flower), *yu* (玉, jade), and *cai* (彩, color). Modern names sometimes are influenced by non-Chinese people, so parents name their babies Mary (玛丽) or Henry (亨利).

Chinese names are not only symbols of social status and etiquette but also signify Chinese culture, social development, and change.

As a special social and linguistic phenomenon, names relate to cultures extensively and profoundly. There are many similarities and differences in Chinese and English names. The authors feel the reasons may be because in ancient times, human beings were not only afraid of the power of nature and natural phenomena, but they also respected them. Therefore, both in Chinese and English personal names, we can find many names derived from places, animals, plants, objects in nature or natural phenomenon." For example, in Chinese names, the Yan Emperor (炎帝, considered an ancestor of the Chinese people) lived near Jiangshui River (姜水), so his clan used Jiang (姜) as their clan name. In English, people used natural phenomenon as family names, such as Frost, Rain, and Snow. Some names come from animal names, for example, Bird, Bull, Lion, Fox, Wolf and so on. Some given names are from plant names and flowers, like Lily and Rose. The authors assert this phenomenon reflects similar cultural connotations in the East and the West that names are associated with primitive totemism. Personal names in Chinese culture follow a number of conventions different from those of personal names in Western cultures. Most noticeably, a Chinese name is written with the family name first and the given name after it.[11] For example, sports enthusiasts around the world are familiar with the basketball player Yao Ming. He should be addressed as Mr. Yao and not Mr. Ming.

Chinese Culture through Language

假名
Pinyin: Jiǎmíng
Literally: A fake name.
Meaning: Pseudonym: false name used when writing.

乳名
Pinyin: Rǔmíng
Literally: Milk name.
Meaning: Infant name, or pet name.

艺名
Pinyin: Yìmíng
Literally: Art name.
Meaning: Stage name.

宗谱
Pinyin: Zōng pǔ
Literally: Family book.
Meaning: Family tree; the record of the family genealogy.

CHAPTER 19

The Chinese Zodiac

十二生肖

In traditional Chinese culture, the twelve zodiac or animal signs were used to record years. Chronologically, they are the rat, ox, tiger, rabbit, dragon, snake, horse, goat, monkey, rooster, dog, and pig. One cycle is twelve years.

There are many legends about why the little rat comes first in the Chinese zodiac. A popular one says that the Yellow Emperor wanted to choose twelve animals as animal signs by a fair competition. Many animals took part in the race. The ox at that time ran very fast and was going to be the first, but the cunning rat who was riding on the ox's back secretly, at the last moment, jumped over the finish line first. As a reward, the god of heaven made it the first animal of the zodiac, with the ox coming in second.

The ox, second in the zodiac, is a symbol of diligence in China. Before farm machines were invented, and even today, the ox helped the peasants to do the hard farm work, so they liked it very much. It is believed that people born in the year of the ox are honest and stubborn and have a strong sense of responsibility and great patience. They also have a positive attitude towards life and care about their families.

The third of the twelve animal signs is the tiger. The tiger is a symbol for dispelling evil and bringing blessings. It is believed that people born in

19.1 A stone-carved Chinese zodiac.

the year of the tiger are confident and often work alone; they have great willpower, are adventurous, and are good leaders. Tigers are a very popular animal among Chinese people. For instance, peasant women like sewing colorful cloth tigers, tiger shoes, and tiger caps as gifts for their relatives and friends. When a new baby is born, a cloth tiger will be given as a lucky gift to wish the child a life as strong and as healthy as that of a tiger.

Another animal sign is the monkey. The monkey is very cute and has been regarded as clever and bright. It is also regarded as auspicious because the sound of the word for monkey is similar to the Chinese word for "nobleman or marquis." In ancient paintings, monkeys played a lucky role. For example, a monkey riding a horse meant someone would become a nobleman immediately. A monkey riding on the back of another monkey suggested the members of the family would be granted titles and land for generations.

The pig is the twelfth animal sign. Chinese people believe that the pig is frank and that people who are born in this year are honest, kind, generous, and very popular. The pig has been a theme in paintings, paper cuts, embroidery, and other folk arts. According to provincial and national news stations during Chinese New Year in 2007, even now more babies are born in the year of the pig, as couples think it an auspicious zodiac year.

Up to now, the Chinese zodiac has greatly influenced the life of Chinese people in various fields. The ancients liked to use clay figurines of the twelve animal signs as burial articles. There were twelve animal sign clocks in Yuanmingyuan (the Summer Palace). Ancient bronze mirrors of the twelve animal signs were made during the Tang Dynasty. They are still popular themes for many Chinese artists. For example, since 1980, China's Ministry of Posts and Telecommunications has issued a set of stamps of the twelve animal

19.2 A festive paper cut of a pig.

signs every twelve years, which are novel and beautiful and well received by many stamp collectors.

Other countries also have animal signs. The twelve animal signs of Mexico are very similar to those of China, including the tiger, rabbit, dragon, monkey, dog, pig, and six other common local animals. Egypt and Greece have the same animal signs: ox, goat, lion, donkey, crab, snake, dog, cat, crocodile, flamingo, ape, and eagle.

Westerners follow the signs of the zodiac (constellations) similar to the Chinese zodiac signs. For example, people who are born between March 21 and April 20 belong to the sign of Aries. This is followed by Taurus, Gemini, Cancer, Leo, Virgo, Libra, Scorpio, Sagittarius, Capricorn, Aquarius, and Pisces. Many Westerners believe that people born under different signs of the zodiac may have different personalities, behavior, and destinies. This is similar to Chinese belief in the zodiac. The Western zodiac is divided into months, and Chinese animal signs into years.

19.3 This painted zodiac also features the four directions in the center.

Chinese Culture through Language

塞翁失马 焉知祸福

Pinyin: Sàiwēngshīmǎ yān zhī huò fú

Literally: A man living in the frontier lost his horse, but it may not be a bad thing.

Meaning: A blessing in disguise.

初生牛犊不怕虎

Pinyin: Chūshēng niúdú bùpà hǔ

Literally: The newborn calves are not afraid of tigers.

Meaning: Young people dare do anything and fear nothing.

打草惊蛇

Pinyin: Dǎcǎojīngshé

Literally: Stir the grass and alert the snake.

Meaning: To act rashly and alert the enemy, to wake a sleeping dog.

画虎类犬

Pinyin: Huà hǔ lèi quǎn

Literally: Draw a tiger that looks like a dog.

Meaning: To make a fool of oneself through excessive ambition.

虎头蛇尾

Pinyin: Hǔtóushéwěi

Literally: Tiger-head-snake-tail.

Meaning: A good beginning with a lousy ending.

狡兔三窟

Pinyin: Jiǎotùsānkū

Literally: A wily hare has three burrows.

Meaning: A crafty person has several ways out of a predicament.

九牛一毛

Pinyin: Jiǔniúyīmáo

Literally: A single hair out of nine ox hides.

Meaning: Something is trivial and unimportant.

鸡犬升天

Pinyin: Jīquǎnshēngtiān

Literally: Chickens-dogs-ascending-heaven.

Meaning: Used to describe anyone who promotes his relatives or friends to high positions simply because he has the clout and power.

老马识途

Pinyin: Lǎo mǎshítú

Literally: Old-horse-knows-the way.

Meaning: A person with experience knows the ropes and has the knowledge of a veteran.

骑虎难下

Pinyin: Qíhǔnánxià

Literally: Ride-tiger-hard-to get off.

Meaning: Someone who is in a difficult situation but still must carry on.

三人成虎

Pinyin: Sān rén chéng hǔ

Literally: Three men make a tiger.

Meaning: A lie, if repeated often enough, will be accepted as truth.

纵虎归山

Pinyin: Zòng hǔ guī shān

Literally: Let the tiger escape; let the tiger return to the mountain.

Meaning: To make any decision which may cause calamity in the future.

CHAPTER 20

Chinese Marriage Customs
中国婚俗文化

China has a long history with various marriage customs. In modern China, various traditional wedding customs and ceremonies have gone out of practice, but recent years have seen a resurgence in traditional wedding ceremonies.

Wedding ceremonies have been considered important throughout human history. "Three Letters and Six Etiquettes" were the basic principles of traditional Chinese marriage customs in ancient times.

Three Letters includes a betrothal letter, a gift letter, and a wedding letter. The betrothal letter is the formal document of the engagement, a requirement for marriage when the bridegroom presents the betrothal gifts as part of the engagement. The gift letter is delivered to the bride's family and is a document necessary before the wedding; it contains lists of the types and quantities of gifts for the wedding. The wedding letter refers to the document that is prepared and presented to the bride's family on the day of the wedding, to confirm and commemorate the formal acceptance of the bride into the bridegroom's family.

Six Etiquettes are proposing (a matchmaker discusses marriage with the potential bridegroom and bride); birthday matching (the matchmaker will ask for the girl's birthday and birth hour, to assure the compatibility of the

20.1 This embroidered wedding garb features the double-happiness character.

potential bride and bridegroom); presenting betrothal gifts (once birthdays match, the bridegroom's family will arrange the matchmaker to present the betrothal letter to the bride's family), the defining of an auspicious wedding day (the groom's family delivers an appointment letter to the bride's family), presenting wedding gifts (after the betrothal letter and betrothal gifts are accepted, the bridegroom's family will formally send wedding gifts to the bride's family, enclosing the gift letter), choosing a wedding date (an astrologist or astrology book will be consulted to select an auspicious date to hold the wedding ceremony), and the wedding ceremony (the bridegroom departs from his home with a troop of escorts and musicians to go to the bride's family and take her back home).[12]

The modern wedding ceremony generally refers to the activities that are held on the official wedding day, but in ancient times Three Letters and Six Etiquettes included the marriage etiquette within the activities of matchmaking, engagement, and wedding ceremony. If the couple did not follow the process, the marriage was not acknowledged. Of these activities, the wedding ceremony is the most elaborate. It is the time when the bridegroom departs from his house and escorts the bride to his home, where the wedding ceremony takes place.

Historical books record the earliest traditional Chinese wedding ceremony:

> In the early morning of the wedding day, the bride puts on her new outfit and a pair of red shoes after a bath in grapefruit water, waiting for a lucky woman to comb her hair. The bride will put on *feng guan* [phoenix coronet] with a red silk bridal veil and listen to a married woman who tells her how to be a good wife, while waiting for her future husband.

20.2 This wedding sedan chair is ready to carry the bride to her destination.

The fun is the groom coming to the bride's door, and the bridesmaids and her sisters trying not to let him in! The purpose is to let the bridegroom experience difficulty when entering the house. Usually, the groom can overcome the difficulty with his wisdom and courage and the help of his friends. Then, the groom sings for the bride, and finally he can see the bride, but the groom must offer red envelopes (hongbao) with money to the bridesmaids and the sisters to be allowed to take his bride home. The bride is accompanied into a sedan by a lucky woman (a middle-aged married woman who will say lucky and auspicious things to her). The bride's sister holds a red umbrella, while another woman sprinkles some rice in the sedan that has a mirror attached, which people think can protect the bride from evil spirits. The bride must cry while departing, showing her unwillingness to leave her parents.

People set off firecrackers to ward off evil spirits while the bride is getting into the sedan. People will try to avoid unlucky things all through the sedan ride: The curtain of the sedan should be closed so the bride cannot see widows, wells, or cats, which are all considered unlucky. Firecrackers greet the groomsmen, and the bride steps on a red cushion to prevent her from walking directly on the floor, which would be bad luck. She must then pass a burning pan to ward off evil spirits before entering the groom's home.

The wedding ceremony is the climax. The bride and groom are shown into the ancestral temple. They first pay respects to heaven and earth, the ancestors and parents, and then each other. After this is completed, they enter the bridal chamber. An emcee narrates the ceremony, accompanied by the wedding guests with applause and cheering. Friends and guests are entertained at a big wedding banquet. The newlyweds toast each other, and drink in thanks to their guests and to receive their blessings.

Traditionally, the newlyweds return to visit the bride's parents three days after the marriage. As part of the tradition, the groom is teased by the bride's relatives and friends; this is out of kindness, of course.

Over time, the traditional ceremony based on Three Letters and Six Etiquettes has gradually been simplified. But wedding ceremony customs are

20.3 Traditional wedding outfits like this are rarely seen in weddings today.

passed down through generations; some traditions have melded into modern trends, whereas others have lost popularity. Sedan chairs, for instance, have mostly been replaced by a car heavily decorated with flowers to greet the bride.

In modern China, traditional wedding ceremonies are not as popular in the cities though they are still in fashion in the countryside. However, in recent times, people have become more and more interested in weddings and are paying much attention to traditional wedding ceremonies.

There are many similarities in marriage customs between China and the West. For example, a Chinese bride throws a ball made of strips of silk to the unmarried girls. The Western bride throws her bouquet of flowers behind her, to the unmarried girls standing there. According to the belief, the woman who catches the ball or the flowers will be the next to get married.

Chinese people are conservative but fun loving, which is fully displayed in the Chinese traditional wedding ceremony. At the end of the wedding ceremony, the friends and family go to the newlyweds' room to celebrate. Chinese believe that the livelier the new home is, the happier the newlyweds will be. In the West, the climax of the wedding is often in the church: The bride's father walks her down the aisle, and then she is handed to the groom by the father putting his daughter's hand into the groom's hands. At the end of the ceremony the priest or minister says: "You may kiss the bride." (This kiss seals the wedding vows they made during the ceremony.) It is then traditional for the bride and groom to have banquet and go on their honeymoon the next day. Many Westerners who believe in God have their wedding in a house of worship. In ancient times Chinese held their wedding ceremonies at home. One of the Chinese traditional marriage customs is *baitian di*; the couple shows respect and thanks to heaven and earth as they become husband and wife. In ancient times, once this was completed, they were considered married.

Chinese Culture through Language

佳偶天成

Pinyin: Jiā'ǒu tiānchéng

Literally: The best partner is made by heaven.

Meaning: Match made in heaven.

父母之命，媒妁之言

Pinyin: Fùmǔ zhī mìng, méishuò zhī yán

Literally: Orders from the parents and words from the go-between.

Meaning: Children must follow the command of parents and the advice of a go-between (in former times); the proper way of contracting a marriage.

情人眼里出西施

Pinyin: Qíngrén yǎn lǐ chū xīshī

Literally: Those in love will see the other as Xi Shi (a famous Chinese beauty).

Meaning: Beauty is in the eye of the beholder.

抛彩球

Pinyin: Pāo cǎi qiú

Literally: Throwing a colorful ball.

Meaning: A way of picking one's mate. In Chinese literary history there are many stories about women throwing a ball to select their future husbands.

指腹婚

Pinyin: Zhǐ fù hūn

Literally: Belly-pointing marriage.

Meaning: A marriage agreement made by parents for the children still in the womb. Pregnant women pointing their abdomens at one another to create a marriage contract. The custom started in the Han Dynasty but is no longer in use.

CHAPTER 21

Chinese Money

中国钱币

China is not only one of the earliest cultures to mint money but also the earliest to issue paper money. China invented many unique minting procedures that influenced many countries and surrounding areas. From shell money to copper coins, silver coins to paper currency, many kinds of money have been used in China over the past four thousand to five thousand years. Different forms of money have also reflected the development and changes of politics, economy, finance, commerce, technology, and culture during different dynasties.

It is hard to say how many kinds of money have been used throughout Chinese history, but it is estimated there have been more than a hundred thousand varieties. Chinese money can be divided into two categories: coin and paper. Coins have been made of different metals such as gold, silver, copper, iron, lead, aluminum, and nickel. Sometimes bamboo, wood, pottery, and porcelain were also used as money. Paper currency includes that printed on cotton paper, mulberry paper, and Daolin paper. People also used currency printed on cloth, such as canvas, sailcloth, and oilcloth.

The earliest currency in China was shells. In China's late New Stone Age (about 10,000 BCE), the barter trade failed to meet the needs of people's social life, so shells were used as a medium for exchange; they were small,

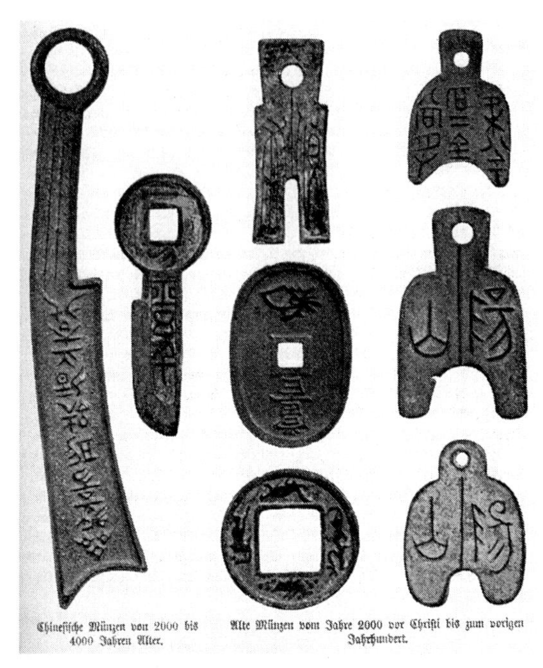

21.1 An assortment of early Chinese coins.

portable, and easy to calculate. Farm tools, spinning wheels, jade, grain, and livestock were also once used as currency.

A great number of coins have been found that were made of lead, silver, gold, and bronze from the Spring and Autumn and Warring States Periods. After Qin Shi Huang, the first emperor, unified China, he unified the written form of Chinese characters, measurements, and money. He banned the use of the old currency that once circulated in the other kingdoms of the Warring States and introduced the new *ban liang qian* (half-liang money), based on the money of the State of Qin. From then on, *ban liang qian* coins became the country's lawfully circulated money.

Because *ban liang qian* was too heavy to carry and inconvenient to use, emperor Wudi of the Han Dynasty founded *wu zhu* coins, which were light enough to use and met the requirements of social economy. Therefore, they were continually produced for many dynasties after the Han. It was the standard currency in circulation for 739 years and the money minted for the longest period.

The first emperor of the Tang Dynasty abolished the old monetary system based on *zhu* and *liang* (units of weight) and issued a new currency called *kai yuan tong bao.* From that point on, metal money was called *tongbao* or *yuanbao*, breaking away from the old way of naming money in accordance with its weight. The decimal system of weight units of *liang*, *qian*, *fen*, and *li* replaced the old unit of *zhu*, which was one-twenty-fourth of one *liang*.

The Song Dynasty saw the peak of China's ancient minting period: A great quantity of many varieties were produced, with superb artistry. This changed in the Yuan Dynasty, when paper notes became the main currency and the use of coins decreased greatly. Although the governments of the Ming and Qing still founded coins with reign titles, each emperor had only one reign title, so there were fewer kinds of coins than during the Song Dynasty.

It is well known that China was the earliest country to use paper currency. During the Tang Dynasty, a kind of paper currency called *feiqian* appeared and could be exchanged for cash in many places. It became the main currency of the Yuan Dynasty and was issued by the government. The paper currency from the Ming Dynasty was called *da ming tong xing bao chao*. In the Qing Dynasty, the main currency was copper coins and silver. Paper currency was

21.2 This silver coin was minted during the Qing dynasty in 1908.

issued in the reign of Xianfeng and was divided into two categories: *da qing bao chao*, which was issued with a denomination based on bronze coins, and *hu bu guan piao*, issued in a denomination based on silver. The last two words of each were combined to create the common word for money *chao piao*, which is still in use in some regions today.

In the middle of the nineteenth century, a great many silver dollars from abroad entered China, and the Chinese government copied them. Since 1949, China has issued many sets of *renminbi* (RMB), the current name of China's currency. Now, Chinese money minting and printing technology are advanced.

For the sake of comparison, the world currency system can be divided into two monetary systems, the Eastern currency system represented by China, and the Western currency system represented by Greece. Eastern currencies were made of copper and iron, characters were used as adornment, and foundry technology was used. Western currencies were made of precious metals, such as gold or silver; figures were used as adornment, and pressing technology was used. However, one kind of coin was made combining the

21.3

technologies of East and West, the famous *han qu erti qian*, a round coin made of copper. The coin was made with a hole in the center and manufactured through a pressing technology which originated in Greece. It has distinctive cultural and geographical features, including a horse pattern on one side surrounded by Kharosthi (an ancient Indic script), and Chinese characters written on the other side. Chinese money today has not only inherited the traditional minting techniques of old but has also learned from Western minting technology. In fact, RMB is a product of the combination of Eastern and Western minting technologies.

Chinese Culture through Language

有钱能使鬼推磨

Pinyin: Yǒu qián néng shǐ guǐ tuī mó

Literally: Having money can make the devil push the millstone.

Meaning: Money makes the mare go.

金字招牌

Pinyin: Jīnzì zhāo pái

Literally: Gold-lettered signboard.

Meaning: Meaning it's a first-class store with a good reputation.

一诺千金

Pinyin: Yī nuò qiān jīn

Literally: One promise, a thousand pieces of gold.

Meaning: A promise that will be kept.

一贫如洗

Pinyin: Yī pín rú xǐ

Literally: As poor as completely washed out.

Meaning: Utterly destitute.

CHAPTER 22

Chinese Food Culture

饮食文化

China has some of the most varied cuisine in the world. Chinese people love eating, and it has become one of China's favorite pastimes. Chinese people have spent a lot of time perfecting cooking and have created regional cuisine and snacks in Beijing, Sichuan, Jiangsu, Guangdong, Shandong, and other places, each with unique characteristics. One can experience Chinese culture through eating Chinese food.

Chinese cuisine includes a variety of different flavors, due to China's vast geography and diverse nationalities.

There are many cooking schools of Chinese cuisine. The most influential and representative provincial cuisines recognized throughout the country are: Shandong, Sichuan, Guangdong, Fujian, Jiangsu, Zhejiang, Hunan, and Anhui. These are often regarded as the Eight Regional Cuisines in China.[13]

Shandong cuisine is mainly composed of Jinan and Jiaodong local food. Jiaodong dishes use stir-frying, deep-frying, frying and steaming, and leavening techniques. The dishes are fresh and somewhat mild. Jinan is known for its soups supplemented by frying, roasting, and deep-fried dishes, which are distinct, fresh, crisp, and tender. Some famous dishes are braised intestines in brown sauce, sweet and sour Yellow River carp, and braised chicken. Shandong cuisine is considered northern. It has played a leading role

22.1 Mapo tofu is a spicy dish composed of soft tofu, minced pork, onion, and Sichuan peppers.

in Chinese cuisine since the Han and Tang Dynasties and remains popular today.

Sichuan cuisine is one of the most famous Chinese cuisines; the spices used are diverse and distinctive. The most famous are the "three peppers": Chinese prickly ash, pepper, and hot pepper; and the three incenses: onions, ginger, garlic. Sichuan cuisine comprises many tastes such as sweet, sour, spicy, bitter, and salty, and eight flavors which are fish flavor, hot and sour, spicy and pungent, *guaiwei* or strange taste, spicy, with chili oil, with ginger juice, and home-style. There are thirty-eight cooking methods and more than three hundred kinds of dishes. Examples include Dengying beef, boiled beef, *fuqi feipian*, mapo tofu, and spiced fish stew.

Guangdong Cuisine or Cantonese cuisine is made up of Guangzhou, Chaozhou, and Dongjiang cuisines. Cantonese cuisine is creative with an emphasis on the artistic presentation of many ingredients and is enjoyed for its light, crisp, and fresh flavor. It is well-known for seafood, soup, vegetarian dishes, and sugar beets. Dongjiang cuisine—Hakka cuisine—is mainly cooked with meat and very little seafood. The ingredients of the dishes are various with a strong fragrance and a heavy, salty taste; the casserole has a unique local flavor. Cantonese cuisine specializes in cooking snakes, raccoon, dogs, cats, monkeys, rats, and other wild animals. Famous specialties include suckling pig, frozen meat, melon cup, and Daliang fried milk. Cantonese cuisine originated from the Western Han Dynasty.

Fujian cuisine is a general term that includes Fuzhou, Xiamen, and Quanzhou cuisine. Common cooking methods are boiling with broth,

22.2 Red chilies and green and black peppercorns spice up most Sichuan dishes.

deep-frying, and stir-frying. Dishes are commonly seasoned with red wine, sugar, and vinegar, and there is an emphasis on making soup. Famous dishes are *fo tiao qiang*, wine-pickled chicken, snowflake chicken, and chrysanthemum fish balls.

Jiangsu (Su) cuisine includes Jiangsu, Nanjing, and Huaiyang dishes. The main cooking methods are stewing, simmering, steaming, roasting, and frying; it emphasizes composition, aesthetics, and beautiful colors. Specialties are Jiaohua chicken (beggar's chicken) lotus leaf chicken, pork ball stew, squirrel-like mandarin, and others. In the Tang and Song Dynasties, Jiangsu cuisine and Zhejiang cuisine were considered two important cuisines of southern China.

Zhejiang cuisine is composed of Hangzhou, Ningbo, and Shaoxing cuisines. Zhejiang cuisine is enjoyed for its freshness, tenderness, and mellow flavor. Hangzhou, Ningbo, and Shaoxing represent three kinds of locally flavored dishes that have been famous since early times. Cooking methods are mainly frying and braising. Ningbo cuisine is fresh and salty, featuring steaming, roasting, stewing. It is known for its seafood, which is tender, soft and slippery, and maintains the original flavor of the seafood. Shaoxing specializes in cooking river food and poultry, which are made crisp and sticky, with a rural flavor. The specialties are West Lake vinegar fish, Dongpo meat, homemade meat, lotus leaf steamed meat, West Lake brasenia soup, Longjing shrimp, and Hangzhou stewed chicken.

22.3 There are many regional recipes for roast duck.

Hunan cuisine is made up of Xiangjiang cuisine, Dongting Lake flavors, and dishes from mountainous areas in western Hunan. The cooking methods are mainly smoking, steaming, and deep-frying. Dishes are characteristically heavy in oil and rich in color. The specialties are Dongan chicken, red simmer shark's fin, steamed preserved meat, breaded whole duck, spicy bamboo shoots, chestnut cooked vegetables, Wuyuan chicken, and spicy chicken, to name a few.

Anhui cuisine features the local culinary arts of Huizhou. It comprises the specialties of south Anhui, Yanjiang, and Huai Bei. The highly distinctive characteristics of Anhui cuisine lie not only in the elaborate choices of cooking materials but also in the strict control of the cooking process. Cooking methods include frying, stewing, steaming, and deep-frying. Anhui cuisine

22.4 Peking duck is famous for its thin, crisp skin. It is served with pancakes, scallions, and hoisin sauce.

22.5 A traditional Chinese kitchen.

started in the Han and Tang Dynasties and became popular in the Ming and Qing Dynasties. The traditional Anhui specialties are ham and turtle, stewed civet cat, Wuwei smoked chicken, Huangshan braised pigeon, and others. The ham snapper stew is also known as turtle-water chestnut consommé.

When entertaining guests, many Chinese think that the richness of the food they order represents their hospitality, and much emphasis is put on the meal itself. The food and wine at a banquet should be rich and colorful, and it must be delicious and expensive. When a feast starts, the host urges guests to eat and drink and serves them frequently by picking up food with his own chopsticks. In the host's opinion, the goal of entertaining is to let guests eat and drink to their heart's content.

Chinese Culture through Language

酒逢知己千杯少，话不投机半句多

Pinyin: Jiǔ féng zhī jǐ qiān bēi shǎo, huà bù tóu jī bàn jù duō

Literally: When drinking with a bosom friend, a thousand shots are too few; when the conversation is not agreeable, half a sentence is too much.

Meaning: When you are with a good friend there is never enough time together.

敬酒不吃吃罚酒

Pinyin: Jìng jiǔ bù chī chī fá jiǔ

Literally: To refuse a toast only to be forced to drink a forfeit.

Meaning: To hesitate to do something until forced to do more.

今朝有酒今朝醉

Pinyin: Jīn zhāo yǒu jiǔ jīn zhāo zuì

Literally: Today we have wine so today we get drunk.

Meaning: Enjoy while you still can.

醉翁之意不在酒

Pinyin: Zuì wēng zhī yì bù zài jiǔ

Literally: The intention of the drunkard is not on the wine.

Meaning: Many kiss the baby for the nurse's sake.

天下无不散的宴席

Pinyin: Tiān xià wú bú sàn de yàn xí

Literally: There is no banquet that does on break up at last.

Meaning: All good things must come to an end.

味如鸡肋

Pinyin: Wèi rú jī lèi

Literally: Tastes like chicken ribs.

Meaning: Describes anything that is of little value but would be a pity to throw away.

因噎废食

Pinyin: Yīn yē fèi shí

Literally: Refuse to eat because of being choked before.

Meaning: To avoid something because of a small risk.

自食其力

Pinyin: Zì shí qí lì

Literally: Feed yourself using your own power.

Meaning: Support oneself by one's own labor.

CHAPTER 23

Chinese Alcohol

酒文化

There are three kinds of ancient alcohol. They are yellow rice wine, wine, and beer. Yellow rice wine was originally made in Shaoxing, China. China has a five thousand-year history of wine-making and has been improving its technology for centuries. In addition to yellow rice wine, there are many different types of alcohol, such as distilled spirits, fruit wine, and medicinal liquor. China has a rich wine and spirits culture.

Chinese alcohol has always been associated with famous poems and great men. The poem "Only Dukang Wine can Allay Misgivings," written by Cao Cao (a famous Chinese warlord and a chancellor of the Eastern Han Dynasty) has enjoyed great popularity for thousands of years, demonstrating Chinese people's long appreciation of wine. Most Chinese alcohol is distilled spirits made from cereal crops. Thousands of years ago, Chinese made a grain alcohol by using special brewing technology. A unique culture was formed from this process.

There is an indissoluble bond between wine and Chinese people. When they are happy, wine is used to celebrate; when they are sad, it can eliminate their gloom; when they are lonely, it can be their best friend. No matter whether they are young or old, of high or low status, people usually become frank and sincere when they drink alcohol. An old saying, "Ale in, truth out,"

23.1 Many Chinese rice wines are made and stored in earthenware containers.

describes this thinking. People tend to reveal their natural, innocent, and pure nature when drinking. The ancient poets and writers believed they would have a greater sense of imagination if they drank, and alcohol often produces unexpected inspiration. It was said that the famous Chinese poet Li Bai, who is known as a saint poet, and Jiu Xian (the god of wine), could not even write poems without drinking some alcohol.

Although wine is not a necessity in daily life, alcohol is used to entertain guests on special occasions such as weddings, births, and birthday parties. It is also used to entertain friends, and it adds to the fun of gatherings, especially during festivals.

Many ethnic minorities have their own drinking customs. Tibetan people like to drink Qingke barley wine. When the guests are toasted, they must show respect to the gods first and then finish the drink in three sips.

Mongolians like to drink horse milk liquor in a big bowl. It is polite to drink the entire glass when someone toasts you. The Yi people in Liangshan, Sichuan Province drink buckwheat or corn wine. In a friendly and harmonious atmosphere, hospitable hosts make sure that their guests can enjoy themselves. Therefore, many ethnic minorities use songs and dance to encourage their guests drink.

23.2 Three popular varieties of Chinese rice wine.

It is worth mentioning that Chinese medicinal liquor, well known in many parts of the world, can improve people's health and cure many diseases.

Every country has its own favorite wine and unique wine culture. France is famous for champagne and wine. In France, people like drinking wine slowly. There are some pairing rules; for example red wine goes with red meat, and white wine goes with fish. The favorite drink for Russians is vodka, a kind of spirit. It is said that before Russians drink it, they first gurgle it in their throats, a tradition handed down from Peter the Great hundreds of years ago. Germany is famous for beer. Germans prefer using tankards to drink beer. Americans make whiskey, known all over the world.

Chinese Culture through Language

酒囊饭袋

Pinyin: Jiǔ náng fàn dài

Literally: Wineskin and rice bag.

Meaning: Someone who can only eat and drink; a good-for-nothing.

酒肉朋友

Pinyin: Jiǔ ròu péng yǒu

Literally: Wine and meat friends.

Meaning: Fair-weather friends.

醉生梦死

Pinyin: Zuì shēng mèng sǐ

Literally: drunk alive, dream to death.

Meaning: Lead a befuddled life as if drunk or in a dream; dream one's life away.

CHAPTER 24

Chinese Tea

茶文化

Tea was first discovered and produced in China. In China, tea is the most famous beverage. The Chinese people have a rich experience in planting and processing tea. Chinese tea is famous worldwide for its high quality and many varieties. All tea-producing countries have imported tea plants and seeds from China directly or indirectly. Hence, the pronunciation of "tea" (cha) in different languages is similar to the pronunciation of tea in Fujian and Guangdong Provinces. Examples include Japanese (cha) and Hindi (chai).

Tea, coffee, and cocoa are the three beverages that enjoy the greatest reputation and consumption throughout the world.

China was the first nation to discover and drink tea as a beverage. There are many stories about the origins of tea. The character for "tea" emerged in approximately the Tang Dynasty. When Lu Yu (the saint of tea) compiled his masterpiece, *The Scripture of Tea*, he decided that it was necessary for tea to have a unified term. At this time the Chinese character 茶 (*cha*) was adopted, and it has been in use ever since.[14] It has been said that tea was found much earlier by Shen Nong, or in the Qin and Han Dynasties.

Chinese tea can be sorted into six types through characteristics of processing and fermentation.

24.1

Green tea is not fermented, and it has the widest regions of production, the largest output, and the finest quality. Among them are some relatively famous types: West Lake Longjing tea, Jiangsu Biluochun tea, Lushan Cloud and Fog tea, Huangshan Maofeng tea, Melon Seed tea, Southern Rain tea, Anhui Green Snow, and Sichuan Bamboo Leaves.[15]

The effects of green tea are superior to the effects of all others. Green tea's anti-cancer functions surpass those of black tea. It is believed that senior citizens should not drink green tea, especially those who are usually constipated, as it may make the condition more serious. But young boys and girls in the development period can drink green tea. Those who are engaged in reading and writing should often drink green tea.

Black tea (called red tea in Chinese) is fully fermented and is the second most famous tea in China. The most famous black teas are Qimen Kungfu, Huhong Kungfu, and Diangong Kongfu. Tea buds are the raw material; they are dried, twisted, oxidized, and fermented indoors, and then dried to produce the final product of black tea. Black tea is good for the stomach and helps urinary output, dieting, and slows aging. Black tea is the best choice for physical laborers and is very good for women after childbirth, especially if consumed with brown sugar.

White tea is only slightly fermented. It has many classifications, such as Great White, Narcissus White, and Young White. Fujian White Hair Acupuncture Needle, White Peony, and Shoumei all belong to the white tea family. It is very rare and valuable; the tea buds are covered with hair as white as snow. White tea is famous for its elegance and has more than 880 years of history.

Yellow tea is also slightly fermented; it is similar to green tea, but it is yellow after it is processed. One can make yellow tea soup; the fragrance is sharp, and the taste is sweet. The best varieties are Mount Jun silver needle and Mongding yellow bud.

Heicha (black tea) is unique and only produced in China. It is a post-fermented tea and is produced from old tealeaves. Varieties include Hunan

24.2 Chinese tea can be loose leaf, pressed into blocks, or made into flowering tea balls.

Heicha, Hubei Old Green tea, Sichuan Biancha, Guangxi Liubao, tea and Yunan Pu'er tea. *Heicha* is famous because of its long history. Generally it is produced in Hunan, Hubei, Sichuan, Yunnan, and Guangxi. Yunnan's Pu'er is especially famous all over the world.

Oolong tea is also called *qingcha*, which is produced by combining the craft of black and green tea production. It is known as a half-fermented tea. The most representative varieties are Oolong tea from the north of Fujian Province, Mount Wu Yi Roch tea, Narcissus, Dahongpao (the Scarlet Gown), Anxi Oolong tea, Guangdong Oolong tea, and others. Oolong tea has the functions of black tea and green tea; it is as fresh as green tea and as clear and delicate as black tea. Oolong tea contains less caffeine, and thus everyone can drink it. It also has the function of producing urine and is a good for weight reduction.

In addition, there are two types of reprocessed tea: pressed tea and scented tea (flower tea). Pressed tea is convenient for long-distance transport and storage. Dry and compressed tea is generally produced from *heicha*, whereas scented tea is made of green tea or black tea with fragrant flowers such as jasmine or magnolia.

The ten most famous teas in China are West Lake Longjing tea, Tie Guan Yin, Huangshan Maojian, Junshan silver needles, Huang Shan Mao Feng, Keemun black tea, Lushan clouds, Xinyangmaojian, Wuyi Yancha, Dongting Biluochun, and Luan seeds.

Chinese tea plays a role in cultivating one's true character, supporting friendship, performing ceremony, cultivating character, and beautifying one's life. Tea has been able to adapt to different sectors and many occasions. The nature of tea is in line with the nature of the simplicity and honesty of Chinese people.

Chinese people's love of tea has created and shaped the Chinese tea ceremony. Chinese tea culture continues to spread to neighboring countries and influence their food culture.

China is an ancient, civilized country; it is a land of ceremony and decorum. Whenever guests pay a visit, it is necessary to make tea and serve it to them. Before serving tea, the host may ask their preference and serve them

24.3 This set includes a tea scoop and other tools such as a tea needle, which prevents the spout from clogging.

the tea in the most appropriate teacups. Snacks, sweets, and other dishes may be served at tea time to complement the fragrance of the tea and to allay one's hunger.

The British are the second largest tea consumers in the world, each person consuming 2.1 kilograms (4.6 pounds) per year on average. Usually black tea is served with milk and sometimes with sugar. For the working class of the United Kingdom, tea break is an essential part of a day. Tea is not only the name of the beverage but also of a light, late afternoon meal, irrespective of the beverage consumed.

Chinese Culture through Language

开门七件事

Pinyin: Kāi mén qī jiàn shì

Literally: Seven things (to consider when) opening the door (to start the day's work).

Meaning: Ancient Chinese believed there were seven daily necessities: firewood, rice, oil, salt, soy sauce, vinegar, and tea.

碧螺春

Pinyin: Bì luó chūn

Literally: Green-spiral shell-spring.

Meaning: A type of green tea originally produced in the Taihu Lake area.

炒青

Pinyin: Chǎo qīng

Literally: Fry-green.

Meaning: A process of curing tea leaves.

珠茶

Pinyin: Zhū chá

Literally: Pearl tea.

Meaning: Pearl green tea. The leaves are rolled into balls and look like pearls.

茶道

Pinyin: Chá dào

Literally: The "Way" of tea.

Meaning: The rituals and processes of making and savoring tea. This refers to the process of steeping tea, and the tools and gestures used in tea making.

CHAPTER 25

Chinese Clothing

中国服饰

Chinese clothing has a long history. The bone sewing needles found in the ancient remains of the caves of Zhoukoudian, Beijing, show that the people of the Paleolithic period learned to make clothes. By the late Neolithic period, people from different regions and nations had designed and worn many different styles of the clothes.

The earliest style of ancient Chinese clothing is the coat and the skirt. During the Shang Dynasty, people began to wear cross-collar garments and skirts on the upper and lower parts of their bodies, loincloths in the middle, and curled–toe shoes. After Emperor Qin Shi Huang united China, clothes showed distinct differences. The working people wore short garments made of coarse cloth[16] called *bu yi*; the rich usually wore long, loose-fitting robes. This time established the basic shapes and designs of Chinese clothing.

In the Spring and Autumn Period and the Warring States Period, the appearance of *shen yi* and *hu fu* clothing styles were the greatest changes in Chinese clothing. At that time people sewed the tunic and skirt together to make a new kind of garment, *shen yi*. It was worn by people from the Warring States Period to the Western Han Dynasty and could be worn by both men and women in all ranks. *Hu fu* was the dress of a minority nationality (a non-Han nationality) living in the north of China. In order to ride horses easily,

25.1 There are many styles of dress among China's ethnic minorities.

they usually wore short garments, trousers, and boots. Because the clothes worn by the Han people were not convenient for working, Wuling, one of kings of the State of Zhao at that time, encouraged his people to wear *hu fu*. He became the earliest clothing reformer in Chinese history. At the same time, Han clothing was enriched and improved.

The policies of recuperating and building the economy during the Han Dynasty strengthened the communication between nations and changed fashion, causing people to dress more extravagantly. For example, *shen yi* worn by women during the Han Dynasty had several layered skirts and many triangular ornaments weighing them down, and long ribbons. After the Eastern Han Dynasty, men no longer wore *shen yi*, but women did. *Shen yi* influenced Chinese women's clothing so greatly that the *qipao* (cheongsam) and one-piece dress evolved from it.

During the Western Zhou period, ranking systems were established gradually and the integrated clothing system for government officials came into being. From then on, there were different color clothes for different people, differentiating the emperor and others with status from commoners.

For example, during the Tang Dynasty, the color of the emperor was yellow, and the color of officials' clothes was differentiated by rank. The first to the ninth rank wore colors such as purple, red, green, blue, black, and so on. Later, no one was allowed to wear yellow or gold clothes. This continued until the Qing Dynasty.

25.2 A Qing dynasty princess poses in her finery.

In ancient China, the gowns worn by emperors were called *long pao* or dragon gowns, which were mainly yellow and embroidered with elegant dragon patterns. For example, the *long pao* of the Qing Dynasty was embroidered with three dragons in the front and three on the back, two on the shoulders, and one inside the front, so there were nine dragons in total. Five can be seen from the front or back. In ancient China, the numbers nine and five were reserved for the emperor, as they symbolized dignity. Other motifs such as spray or spindrift, hills and treasures, would be embroidered at the lower hem of the gown. These patterns represent good fortune and the people's wish for the emperor to keep the country united and peaceful forever.

Men's clothing in the Qing Dynasty was the gown and *magua*. *Magua*, also known as the mandarin jacket, was usually worn over the gown, which was originally the riding clothes of the Manchu. After the Xinhai revolution, although people still wore gowns and a mandarin jacket, Sun Yat-sen's uniform gained popularity and gradually became the main style for men. (Sun Yat-sen was a great reformer and considered the father of the Chinese nation.) It has the combined advantages of Western style, Chinese style, and Japanese students' uniforms. Dr. Sun liked this uniform very much and was the first one to wear it, so it was named after him.

25.3 Only the emperor was allowed to wear long pao. made of the finest silk.

25.4 Chinese beauties wearing qipaos.

The *qipao* is an elegant type of Chinese dress. This tight-fitting dress comes from the Manchu ethnicity. It was so long that it could reach a woman's insteps. Later, the *qipao* became popular throughout China and underwent many changes. In fact, it is said that the *qipao* is a combination of many cultures including Manchu, Qidan, Mongolian, and Han. Nowadays, the *qipao* is still one of the first choices for many Chinese women when they attend important social gatherings.

Similar to the evolution of Chinese clothing, Western-style clothes evolved from daily life and work. It is said that the modern-day suit was originally fishers' clothing; it was worn at sea all year round and was very convenient for fishing work because of its open-necked collar and minimal buttons. The daintiest, swallow-tailed coats evolved from coach driver's clothing from medieval Europe.[17]

Chinese Culture through Language

天衣无缝

Pinyin: Tiān yī wú fèng

Literally: A seamless heavenly robe.

Meaning: Flawless; perfect; without a trace.

衣锦还乡

Pinyin: Yī jǐn huán xiāng

Literally: Return to one's hometown in silken robes.

Meaning: To return home after making good.

衣冠禽兽

Pinyin: Yī guān qín shòu

Literally: A beast in human clothing.

Meaning: A brute; a dressed-up beast.

衣不解带

Pinyin: Yī bù jiě dài

Literally: Not to undress.

Meaning: Sleep without taking off one's clothes; depicts someone who has worked so hard they slept in their clothes.

衣冠楚楚

Pinyin: Yī guān chǔ chu

Literally: Clothes and the hat are all tidy.

Meaning: Neatly and immaculately dressed.

CHAPTER 26

Chinese Embroidery

中国刺绣

Embroidery is one of the outstanding traditional crafts of Chinese people. Shangshu (the book of history) recorded as early as four thousand years ago that "clothing should be designed and embroidered." As a result, embroidered clothes became widely fashionable. Today, embroidery is very popular in the whole country and embellishes many household products like quilt covers, pillows, and cushions.

Embroidery is a craft that deals with different kinds of fabric and various materials woven into different patterns. From the magnificent dragon robe worn by emperors to the popular embroidery seen in today's fashions, embroidery adds much beauty to Chinese life and culture. It also has an influence on international cultural life.

Chinese embroidery is a long-established art form in China and has many categories: hand silk thread embroidery, metallic yarn embroidery, beaded embroidery, cross-stitch, and appliqué. The oldest embroidered product from China on record dates from the Shang Dynasty, which reached its peak during the Spring and Autumn Period. Embroidery in this period symbolized social status, as it did later during the Qin and Han Dynasties. However, it was not until the development of a national economy that embroidery entered the lives of the common people.

Though Chinese embroidery has a long history, it was never classified as a solely female activity; men and women, old and the young, have been involved in embroidery. The items embroidered are quite diverse: clothes, theatrical costumes, tablecloths, pillowcases, cushions, screens, and wall hangings.

According to the stitches used, there are two main divisions of embroidery; the long stitch and the short stitch, and the seed stitch used in Beijing, also known as the French knot. The stitches most commonly used by the Chinese are: 1) satin stitch, which is further classified into long and short; 2) Beijing stitch or French knot; 3) stem stitch; 4) couching; 5) chain stitch; and 6) split stitch. All these stitches are known in the West. Many Westerners find Chinese embroidery very ornate. When done perfectly, the Chinese satin stitch is exquisite because of its fine detail. In Chinese embroidery, the four major traditional styles are Su, Shu, Xiang, and Yue.

There are many kinds of embroidery products in Suzhou, referred to as Su embroidery. Suzhou is a typical southern water town, renowned for its silk. An old saying goes, "Every family breeds silkworms and every household does embroidery." As early as the Song Dynasty, Su embroidery was well known for its elegance and vividness. It is said that, during the Three States Period, Zhao Wufei, the wife of the emperor Da of Wu (Sun Quan), excelled in embroidery. Her embroidery work, "the map of states," was praised and called "the 10,000 states embroidered in brocade." Because of this she gained fame as the "matchless needle." Later, in the Qing Dynasty, Su embroidery products were widely used in clothing, pillows,

26.1 Embroidered slippers.

cushions, and other daily necessities. Su embroidery is famous for its beautiful patterns, elegant colors, variety of stitches, and consummate artistry. Its stitching is meticulously skillful, and its coloration subtle and refined.

Historically, Su embroidery dominated the royal wardrobe and even today, it can be found in a large share of the embroidery markets in China and all over the world.

Shu embroidery refers to the embroidery from the west of Sichuan. Shu embroidery, influenced by its geographic environment and local customs, is characterized by a refined and lively style. The earliest record of Shu embroidery is during the Western Han Dynasty. At that time, embroidery was a luxury enjoyed only by the royal family and was strictly controlled by the government.

In the Qing Dynasty, Shu embroidery entered the market and an industry was formed. The composition of this embroidery, which can include scenes from paintings by masters, patterns by designers, landscapes, flowers and birds, dragons and phoenixes, tiles and ancient coins, is diverse and colorful. Legends like *The Eight Immortals Crossing the Sea* and other auspicious patterns such as magpies on plums and mandarin ducks playing on the water were also favorite subjects. Patterns with strong local features were very popular among non-Chinese people at that time. These local features included lotus and carp, bamboo forests, and pandas. Shu embroidery is famous for its smooth lines, rigorous stitches, soft hues, and artistic feel.

Xiang embroidery refers the embroidery products from central Changsha and Hunan regions. It developed from ancient Hunan folk embroidery, and the technique reached maturity during the Warring States period (approximately 475 to 221 BC).

Over two thousand years ago, Xiang embroidery became a special branch of the local art. In addition to the common themes seen in other styles of embroidery, Xiang embroidery absorbed elements from calligraphy, painting, and inscription. Its sketches take the place of traditional painting drafts, retaining the traditional charm of Chinese painting while displaying the elaborate stitches and the intense color of Xiang embroidery. This style of embroidery uses a needle as a pen and thread as the ink. That

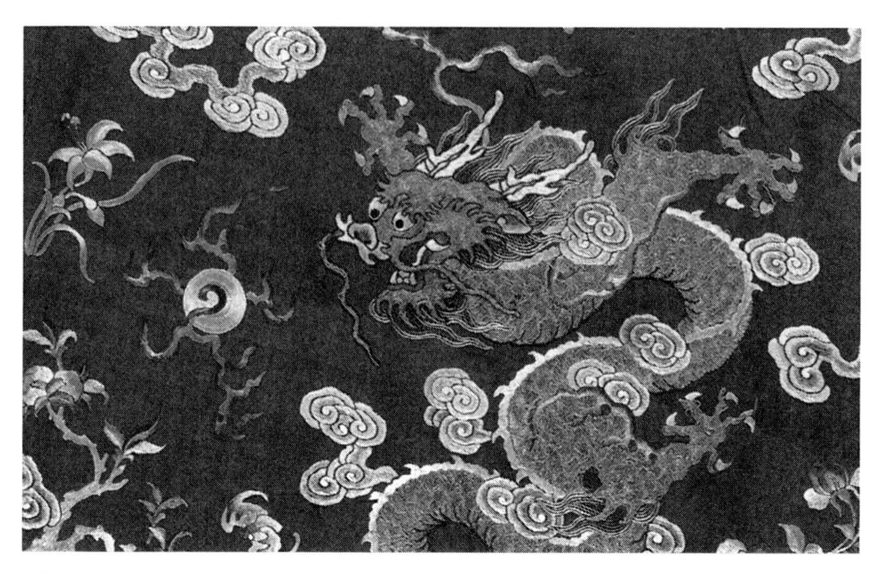

26.2

is why Xiang embroidery is widely appealing and long lasting. In Xiang embroidery, a flower seems to emit fragrance, a bird seems to sing, a tiger seems to run, and a person seems to breathe. Everything comes alive in Xiang embroidery.

Yue embroidery, which encompasses Guangzhou embroidery and Chaozhou embroidery, has the same origin as Li brocade. Portraits, flowers, and birds are the most popular themes, as the subtropical climate produces areas with an abundance of plants that are rarely seen in central China. In addition, Yue embroidery tends to use dark colors; red and green are widely used to strongly bring out a spectacular and lively atmosphere. The composition of Yue embroidery is complicated and intertwined. It includes splendid color, changeable stitches, clear texture, and a strongly decorative aesthetic. Another embroidery form, Naxi embroidery, uses gold embroidery and leather for lining, and its products appear golden and brilliant. The most representative motif of Yue style embroidery is "A Hundred Birds Worshiping the Phoenix," a motif with rich cultural connotations.

In addition to the four famous Chinese embroidery styles, there is Jing embroidery of Beijing, Gu embroidery in Shanghai, Miao embroidery of the Miao nationality, and so forth. Produced in different states, they all have their own particular styles. Nowadays, Chinese embroidery is all over the world. It has become one of China's strongest exports and is very popular everywhere.

The random stitch was created by the embroiderer Yang Shouyu in the 1930s. This stitch combines Chinese traditional embroidery technology with Western art characteristics, which is a new art. According to the light and color changes of Western oil painting, the random stitch recreates this effect through rich and changeable needlework, showing the embroidery's beautiful lines and creating a three-dimensional sense of reality. Random stitch embroidery looks like an oil painting, but it is even more gorgeous because of its bright color; it is a unique style unto itself. Because of this, the random stitch is widely used in embroidering oil painting, photography, and sketches.

Cross-stitch originated in Europe and is a craft with a long history. In 4 CE, it came from Turkey to Italy and then to Europe. Cross-stitch was initially popular in the court. The common people quickly became interested, and later it spread all over Europe, the US, Asia, and other parts of the world. It was the favorite pastime of European women in the 1930s and is seeing a revival today. In recent years, cross-stitch was introduced into China. People of different ages love it because it is easy to learn.

Chinese Culture through Language

男人看田边，女人看花边

Pinyin: Nán rén kàn tián biān, nǚ rén kàn huā biān

Literally: The man looks at the edge of a paddy field, while the woman looks at lace.

Meaning: This is used to describe the women of the Miao ethnic minority who treasure beautiful clothing. Their clothing contains hand-stitched scenes of the best moments of life and of times gone by. In the past they only put on their splendid attires twice, once during the wedding ceremony, and the other when being buried.

鸳鸯绣

Pinyin: Yuān yāng xiù

Literally: Mandarin duck embroidery.

Meaning: Mandarin ducks appear frequently in Chinese ancient literary work. Items with mandarin ducks are frequently given to the newlyweds to represent eternal love.

百鸟朝凤

Pinyin: Bǎi niǎo cháo fèng

Literally: A hundred birds paying homage to a phoenix.

Meaning: In ancient times, the phoenix symbolized the ruler. Therefore "A Hundred Birds paying homage to a phoenix" shows the common people's wish for prosperity, and expresses their hope for peace.

CHAPTER 27

Chinese Furniture

中国家具

Being an important part of the treasure house of Chinese art and culture, furniture not only has a very long history but also contains unique national characteristics and cultural connotations. Chinese furniture interests many people, especially those who love traditional Chinese culture and enjoy tranquil environments.

Chinese furniture, which is closely associated with social ideology, customs, lifestyles, and aesthetic ideas in different periods of Chinese history, generated from mats that were used to sit on the ground, and then gradually developed into furniture with feet that could be set on the ground (like a chair). Chinese scholars divide Chinese furniture into four groups based on different designs: Chu-style furniture (from the Zhou Dynasty to the Southern and Northern Dynasties), Song-style furniture (from the Sui and Tang Dynasties to the early Ming Dynasty), Ming-style furniture (from the middle Ming Dynasty to the early Qing Dynasty), and Qing-style furniture (after the middle Qing Dynasty).

Chu-style furniture was developed from low types to high types, with Chu cultural characteristics, and the style gradually matured as it developed. *Zu* (an ancient term for objects that hold sacred materials) and *jin* made of bronze, are representative furniture from the Shang and Zhou Dynasties.

27.1 Stone furniture can be found in gardens and parks throughout China.

They were used to make sacrifices to heaven and earth for favorable weather and bumper grain harvests, and reflect characteristics of the times and level of technology. Then, in the Spring and Autumn and the Warring States Periods, tables, screens, and clothing racks were made. Lacquer was very popular at that time, and lacquer furniture, brightly colored and simply gorgeous, was mainly made from *heidi*, with red-colored patterns, and reached the height of its popularity in the Qin and Han Dynasties. As society developed, various low-type furniture gradually appeared, such as beds, couches, tables, desks, cupboards, chests, and clothing racks. In the Wei, Jin, Southern, and Northern Dynasties, people's mode of living changed; high-type furniture was made, such as chairs and square stools. Lacquer furniture was then the most popular style and was the main representative of Chu-style furniture.

The appearance and growth of Song-style furniture was deeply influenced by social politics, economy, and cultural consciousness. Before the Song Dynasty, there were both low-type and high-type furniture; their styles were light and simple, elegant and straight, and not completely rounded. During the Song Dynasty, the varieties of furniture increased, and the types became more beautiful. People went from sitting on the ground to using furniture with arms and legs during this time. High-type furniture became the main form of furniture in people's lives. Such furniture includes desks and chairs, drawers and cupboards, dressing tables, hangers, and towel rails. The structure of most Song-style furniture is simple, and its decorative style elegant, which represents people's aesthetic values in the Song Dynasty, thrifty and neat. The Song Dynasty, which is generally considered the most important development period linking the past and future history of Chinese furniture, established the foundation for Ming-style and Qing-style furniture.

Ming-style furniture, which was based on traditional Song-style furniture, used imported high-quality timber (such as padauk wood, rosewood, cassia, and siamea) to produce hardwood furniture that was used in halls, bedrooms, pavilions, and study rooms. The concept of creating sets of furniture began at this time. Ming-style furniture developed from the middle Ming Dynasty to the early Qing Dynasty, the golden age of Chinese classical furniture. The style of furniture was neat and pleasingly simple; its style was modest, lines were soft and straight, with proper proportion, varied decoration, precise structure, and high-quality materials: It strongly reflected the national style. It combined materials, production, and aesthetic so skillfully that it achieved a high integration of artistry and practicality. The book *The Development of Furniture Home and Abroad* made the following comments about it: Ming-style furniture marked the high period of accomplishment of Chinese classical furniture. Chinese Confucianism, which is broad and profound, ran through its development process, imbuing it with strong characteristics of the times and the national style. It was the pride of the nation, a brilliant achievement in the history of world furniture, and played an important role in the development of world furniture.[18]

Qing-style furniture inherited style characteristics of Ming-style furniture, but its form and tone varies widely from the Ming style. Provided with favorable conditions for furniture production by the Kong and Qian eras, Qing-style furniture used various consummate technologies, mainly with thick and heavy designs, enormous contours, and complicated adornment. Two major characteristics of the Qing style are its good use of materials

27.2 These Ming-style chairs feature stone back panels.

and its varied decoration with high-level artistry. In addition, its decoration techniques included carving, inlays, and lacquer painting; an example is the wooden armchair. In the late Qing Dynasty, folk furniture began to emerge, and three furniture systems with strong local characteristics were formed, from Beijing, Suzhou, and Guangzhou. Beijing furniture directly inherited stylistic aspects of the court living style during the Qing Dynasty but was not as ornate and stylized as court furniture. Suzhou furniture, produced in Suzhou, mainly displayed the characteristics of Ming-style furniture but varied slightly in style; it was considered a representative of Ming-style furniture. Guangzhou furniture, as the representative of Qing-style furniture, maintained the Chinese traditional style. Its variety increased gradually, and the carving of series works was complicated and influenced by the West.

The above four furniture design styles show the course and heritage of Chinese classical furniture, chiefly reflecting the thoughts of and harmony between humans and nature. They are an integral part of the inheritance and promotion of Chinese traditional culture and furniture design.

27.3 An intricately carved Chinese sofa featuring dragons and lions.

Even though each type of art has its own characteristics, the furniture of both China and that of Western countries is gorgeous. In decoration, lacquer painting is characteristically Chinese, whereas carving is characteristically Western. Chinese furniture was influenced by Confucianism, Daoism, Zen Buddhism, rules of propriety, and other schools of Chinese thought. According to Li Minxiu's *Comparative Study of Chinese and Western Furniture Culture*, Western furniture was influenced by pragmatism, authoritarianism, mysticism, humanism, idealism, and other schools of Western thought. Although there are many differences between Chinese furniture and Western furniture, the development of both was dependent on the requirements of material living standards, the context of social life, the pursuit of art, and the style of the time.

CHAPTER 28

Traditional Folk Arts

民间工艺

Chinese folk arts and crafts illustrate Chinese people's aesthetic standards. The popular and practical crafts reflect not only Chinese people's diligence and wisdom but a strong national culture as well.

Chinese folk arts and crafts are colorful and very rich in content with a long history. Most are related to folk tales, the contents of which are based on having lively spirits, good luck, a long life, wealth, and happy reunions with children and grandchildren. Folk arts and crafts are also part of folk festivals and are traditional religious and ethnic decorations. They are lively and interesting, widely known, and represent the Chinese people's traditional culture and daily activities.

Shadow play, a Chinese folk art form, is a unique and artistic combination of art and drama, which shines like a delicate pearl among Chinese arts. It is a folk art originating from rural areas of northwest China's Gansu, Shanxi, and Ningxia Provinces. This artistic form was popular between the fourteenth and nineteenth centuries, during the Ming and Qing Dynasties. Figures used in the play look pretty and charming, and the carving techniques require a high level of skill. The material used for shadow play puppets is ox hide, which is durable, soft, and just the right thickness. People first clean and dry

28.1 A puppeteer must use one hand to move the multiple joints of these life-like figures.

the hide, then draw the outline of the puppet on the hide, and cut out the shape. Then they put it in water to add color to it. After the dying process they iron it, which is the most difficult yet important step. Finally, the hide is dried and bound together for the performance.

The main colors of the leather silhouette are red, yellow, black, and green, yet the figure can display more subtle colors with complicated patterns. Simple figures and exquisite artisanry are its two features. Lines are drawn to depict the general figure. Most parts of the figures are pierced; the non-pierced parts are complementary to each other. Different patterns are used for different parts. The parts of the body composition of shadow puppets are flexible, portable, and easy to manipulate for performances. Their performances include rich and beautiful singing and exciting movement. Shadow plays emphasize the performance. Every part of the body is moved, and each movement is matched with inspiring strains of songs. The whole play displays strong local characteristics.

Shadow play has also played a role in cultural development in China and abroad. There are many new local Chinese operas that were derived from the traditional forms of shadow singing. For thousands of years, shadow plays have been widespread and deeply loved by Chinese people.

Clay figure modeling involves kneading clay to create various figures, animals, flowers, and plants out of clay. Chinese clay crafts are unique in their style because they are beautifully shaped, vivid, and lively. The history of painted clay crafts goes back to the Ming Dynasty more than six hundred years ago. Fengxiang in Shanxi is the birthplace of clay modeling. The local unique white earth was used to make earthen bricks, and the traditional folk handicraft was adopted in the production process of clay sculpture. In the past, farmers enjoyed it in their spare time or at night during busy harvest times. There are ten steps in making clay figures, some of which are molding the clay, putting it into the designed shape, polishing, and painting. Clay crafts enjoyed wide acclaim during the Qing Dynasty, mainly because of the painted clay work done by Clay Figurine Zhang in Tianjin. This artist improved the traditional art by adding colors and props, creating a unique style. Chinese clay crafts have been to World Expos many times, received a number of awards, and are spoken highly of by people in China and abroad. The main subjects in clay art include tigers, sitting cows, zodiac animals, toys, legendary figures like the Eight Immortals, and characters from the Three Kingdoms, especially "little pandas inviting the guests," which are the best gift for children.

Kites, also called *fengzheng*, have a history of more than 2,500 years. They were known in ancient China as *yao* (wooden kites) in the south, and as *yuan* (paper kites) in the north. Kites are one of the well-known handicrafts that show national characteristics. The earliest kite was made by the philosopher Mo Di (ca. 470–ca. 391 BCE), and soon thereafter kites were used in war. During the Tang and Song Dynasties, kites were made of paper, quickly gained popularity, and flying kites became a favorite pastime among common people. It is said that kite flying could help get rid of disaster; people still believe that flying kites can bring good luck and make their crops rich and tall. Kite flying is widespread in China, especially in Beijing and Weifang, both important places of kite manufacturing and

28.2 Chinese opera faces are featured on this billowing kite.

flying. As for the variety of kites, there are soft-winged, hard-winged, dragon-shaped and so on, with more than one hundred themes. The kites manufactured in China have swept all over the world with their exquisite beauty and unique style, occupying an important position in world history, science, and culture.

Chinese folk handicrafts are various and colorful. In addition to the above, China produces cloth, wood carvings, lanterns, pendants and hanging decorations, lacquer, bamboo crafts, jade crafts, porcelain, paper cuts, and New Year pictures. These folk arts are the outcomes of thousands of years of wisdom and Chinese culture.

Each nation has its own particular crafts. For instance, the clogs of the Netherlands, which were traditionally handmade, are the symbol of national culture and custom; Russian Matryoshka dolls (nesting dolls) are special wooden toys with various regional styles. They are the handicraft unique to Russian people. Chinese folk arts and crafts are artistic forms inherited from

28.3 Ethnic minority women use a traditional loom.

regional or ethnic scenes in China. Usually there are some variations among provinces. Individual folk arts have a long history, and many traditions are still practiced today. Chinese folk art is an important part of the country's rich cultural and art heritage. Chinese folk art has won recognition and praise from experts both in China and abroad for its great variety, genuine rural content, rich flavor of life, distinctive local style, and its artistic approaches of romanticism.

Chinese Culture through Language

泥菩萨过江

Pinyin: Ní pú sà guò jiāng

Literally: Clay Bodhisattva crosses the river.

Meaning: Unable to help or save oneself, let along others.

泥塑木雕

Pinyin: Ní sù mù diāo

Literally: A clay or a wooden statue.

Meaning: As wooden as a dummy.

彩陶

Pinyin: Căi táo

Literally: Colorful pottery.

Meaning: Painted pottery: The history of painted pottery goes back to the Ming Dynasty, 600 years ago.

CHAPTER 29

Chinese Jade

玉文化

The Chinese people deeply admire jade; thus, China has been called "the country of jade." In fact, China is the first nation to discover and make use of it. Eight thousand years ago, Chinese people began to make ornaments with jade. They believe that jade is a mysterious and beautiful stone containing the spirit of both heaven and earth.

In China, jade is very popular and is a beautiful, sacred, precious, and auspicious symbol. The Chinese even believe that jade has the five virtues, like humans. The five virtues of jade are the hard texture; smooth, light, beautiful color; dense structure; and pleasant sound that can be compared with benevolence, righteousness, wisdom, bravery, and purity, the five cardinal virtues of humans as laid out in Confucianism. Jade culture has had a profound influence on the spirit of the Chinese nation. In ancient times, jade was also a symbol of hierarchy. In civil society, it was said that jade could bring luck, repel evil spirits, and prevent diseases. The Chinese people have worn beautiful jade for thousands of years. Their love of jade was demonstrated by their creation of medals inlaid with jade for the Beijing 2008 Olympic Games.

Eight thousand years ago, in the Neolithic period, Chinese carved jade into semi-circular jade pendants, pearl-like beads, and other decorations.[19] The jade articles of the Liang Chu culture in Jiangsu and Zhejiang Provinces

29.1 Dragon carved of jade.

represent the highest level of primitive Chinese jade sculpture. At that time, it was common for the nobility to use jade objects as burial articles. An archeological dig at the Fanshan grave site in Hangyu, Zhejiang, unearthed more than seven hundred jade articles in one grave, among which were more than sixty kinds of articles such as jade tablets, bracelets, fish, birds, and turtles. These jade articles are of excellent artisanry.

In the Xia Dynasty, many high-quality jade articles evolved from weapons and tools, such as the dagger axe, knives, battle axe, and others. The jade objects of the Shang Dynasty are represented by those unearthed from the tomb of Fu Hao. There were 755 jade articles in her tomb; among them a fine, kneeling jade figure meticulously carved with a clear facial expression, dress, and hairstyle.

During the Warring States Period, nobility liked the famous *hetian* jade, and pendants made from this jade were very popular at that time. The best jade articles were produced in the Yuan, Ming, and Qing Dynasties. The biggest jade article (an artificial mountain carving), "Da Yu Led People in Curbing Floods," displayed in the Palace Museum in Beijing, is more than two meters high and ten thousand kilograms. It was finished in the fifty-third year of the Qianlong Period (1774, in the Qing Dynasty).

From ancient times, jade has been regarded as precious, so when emperors and nobles died they were usually dressed in jade clothes known as a jade burial suit, sewn with fine gold thread. The one unearthed from the tomb of Liu Sheng, the king of Zhongshan, was composed of 2,498 pieces of jade and 1,100 grams of gold thread. According to the hierarchy of the day, only emperors could wear jade clothes sewn with gold thread; princes and princesses wore those sewn with silver thread; and ones sewn with copper thread were worn by other nobles of lower status.

29.2 Commonly thought of as being green in color, jade can also be orange or white.

The most famous legend about jade is of the He Family Tablet. During the Spring and Autumn Period there was a man called Bianhe, who found an uncut stone that he knew was a rare piece of precious jade. But when he twice presented this piece of uncut jade to the kings, he was punished by having his feet cut off because the jade carvers in the palace said he had cheated the kings with an ordinary stone. At last, he sent the stone to a third king who believed him and let his workers open it, which revealed a fine, priceless piece of jade.

Later, many other kings wanted to possess this unparalleled jade. There is a famous story called "Bring back the Jade Intact to the State of Zhao." In exchange for the precious stone, the king of the State of Qin said he was willing to offer fifteen cities to the king of the State of Zhao, the owner of the jade at that time. But after the king of Qin accepted the jade, he refused to hand over the fifteen cities he had promised, so Li Xiangru, the emissary of Zhao, got the jade back in a clever way. However, when the king of Qin unified all other states of China, he possessed the jade at last and he turned it into his imperial seal.

Chinese Culture through Language

金口玉言

Pinyin: Jīn kǒu yù yán

Literally: Golden mouth and jade words.

Meaning: Precious words that carry great weight.

金玉良言

Pinyin: Jīn yù liáng yán

Literally: Gold and silver good sayings.

Meaning: Good advice.

抛砖引玉

Pinyin: Pāo zhuān yǐn yù

Literally: Cast a brick to attract jade.

Meaning: To offer introductory remarks or commonplace words so others can provide more valuable opinions or comments.

完璧归赵

Pinyin: Wánbìguīzhào

Literally: To return the jade intact to the State of Zhao.

Meaning: To return something to its owner intact.

瑕不掩瑜

Pinyin: Xiá bù yǎn yú

Literally: A speck on a piece of jade won't obscure its beauty.

Meaning: A small flaw will not affect the fine quality of a person or an object.

玉石俱焚

Pinyin: Yù shí jù fén

Literally: Both jade and stone burn together.

Meaning: Destruction of good and bad alike.

玉叶金枝

Pinyin: Yù yè jīn zhī

Literally: Jade leaves and gold branches.

Meaning: Describes persons of imperial lineage.

CHAPTER 30

Chinese Porcelain

中国瓷器

The invention of porcelain is another great contribution of the Chinese nation to world civilization. China is the home of porcelain. In English, porcelain has become a synonym of the word "china". Early Chinese porcelain came into being in the mid-sixteenth century BCE, and porcelain production technology and artistic creativity reached a high degree of maturity in the Tang Dynasty. There was a porcelain boom in the Song Dynasty, and famous kilns emerged. The technology was further improved in the Ming and Qing Dynasties. China's ceramic industry has flourished and is popular among people all over the world, due to its high quality and elegant shape.

Throughout history, porcelain, like silk, with its distinctive ethnic characteristics, serves as a unique contribution to world culture. From the word "china" (meaning porcelain) and China, we can say that, in the eyes of many non-Chinese people, porcelain is associated with China as a country.

The development of Chinese porcelain evolved from pottery and the primitive porcelain that originated over three thousand years ago. Up until the Song Dynasty, famous porcelain and kilns had earned their reputation in over half of China, and this became the most prosperous period. The development of Chinese porcelain can be divided into three stages: celadon, white porcelain, and painted china. The period of celadon dates from the

Warring States to the Yuan Dynasty. Then, in the period of Wei Jin and the Southern and Northern times (220 AD–598 AD), porcelain technology was enhanced, and in the Northern Dynasties (439 AD–534 AD) white porcelain appeared, which could be seen as a marvelous invention in Chinese porcelain history. During the late Northern Period, painted porcelain appeared. It was during the Tang Dynasty that Chinese porcelain reached its peak.

Celadon is fired by applying a green-glaze material (iron is the coloring agent of the blue-green glaze) on the object. Having delicate texture, bright lines, smooth, dignified, but simple shapes and pure color, celadon is famous all over the world. "Green jade, bright as a mirror, sounds like a chime" and "porcelain flower" are titles that made celadon a worthy porcelain treasure. In those days, celadon had the reputation of "surpassing porcelain and being similar to jade."[20]

Based on celadon, white porcelain was eventually produced. During the Tang Dynasty, the technology of producing white porcelain was very advanced; thus many kilns in the northern parts of the country largely baked white porcelain. The white porcelain from the Xing kiln competed

30.1 A variety of Chinese porcelain painting techniques.

with celadon, of which the southern region was the representative, so people said "south green north white." The white porcelain produced in the Tang Dynasty marked the true maturity of the art, its glazed white skin resembling silver and snow. The emergence of white porcelain paved the way for blue and white porcelain and painted china.

Painted china had designs painted on its surface, and its popularity reached a peak in the Tang Dynasty. The shift from the Ming to the Qing Dynasty was a prosperous period for the development of painted China. In this period, the most outstanding achievement undoubtedly is the Jingde kiln. In addition, during the Song Dynasty at the porcelain capital of Jingde, the kilns there produced a kind of porcelain that was white but in which one could perceive the shadow of green (also known as blue porcelain). The blue-green, white, and the glazed red porcelains were known for their fresh pastel coloring. In the Ming and the Qing Dynasties, painted porcelain made unprecedented progress in its purity, texture, and diverse glaze color. In the Qing Dynasty, colored enamel appeared and pink was produced.

30.2

Porcelain serves as one of China's unique commodities and commands a vast market overseas. Since the Sui and Tang Dynasties, China has exported porcelain to countries in Asia, the Middle East, Europe, and the Americas. Even in Sudan, part of Africa, one can find traces of Chinese porcelain. Beginning in the Qing Dynasty (1684), porcelain began to be exported on a large scale. During the reigns of the emperors Kangxi, Yongzheng, and Qianlong, porcelain manufacturing technology became very advanced and produced high-quality and colorful items. Many Europeans came to China to trade for these products.[21] Later, with the permission of the Qing government, trade organizations and agreements were established

30.3

with countries such as Great Britain, France, the Netherlands, and others. Because of the distance and the inconvenience of transportation, those who could afford Chinese porcelain treasured it and eventually some pieces have been put into museums. If you go into any European royal palace, museum, or small country castle, or even into a French friend's house, you can find some Chinese porcelain.

Since the Tang and Song Dynasties, porcelain has become one of China's bulk export commodities, and now many countries have collected porcelain from China.

China has always been considered the birthplace of ceramics; therefore, the ceramic culture of the West is related to China to some extent. A great deal of exchange took place in the eighteenth century. The fusion of Chinese and Western

30.4 Hand painting porcelain requires great focus and skill.

porcelain is called Rococo-style porcelain. During the reign of Emperor Qianlong the artistic features of Chinese porcelain were influenced by the Rococo artistic style, making the patterns in the porcelain more delicate and attractive. A Western scholar once said: "All the secrets of the combination between Rococo art and ancient Chinese culture lie in the delicate porcelain, embodied in its subtle tone and emotion."[22] The famous German scholar Adolf Reichwein once commented that Chinese porcelain is considered a symbol of the Rococo era porcelain because of its unique luster, hue, and textural beauty. The smooth texture of Chinese porcelain, Chinese-style painting patterns, elegant colors, smooth lines, and freehand decoration are similar to the Rococo art style. It uses "S-" and "C"-shaped curves, and emphasizes fine textures and light colors.[23]

Chinese Culture through Language

瓮中之鳖

Pinyin: Wèng zhōng zhī biē

Literally: A turtle in a jar.

Meaning: Trapped.

瓷饭碗

Pinyin: Cí fàn wǎn

Literally: Porcelain rice bowl.

Meaning: Insecure position or job.

套瓷

Pinyin: Tào cí

Literally: Coverable porcelain.

Meaning: Try to establish a relationship with someone; cotton up to.

瓷婚

Pinyin: Cí hūn

Literally: Porcelain wedding.

Meaning: 20th wedding anniversary.

CHAPTER 31

Paper Cutting
剪纸

Paper cutting is a traditional folk art. Created by clever peasant women using skillful and simple techniques, the art of paper cutting has produced works based on people, auspicious animals, beautiful flowers, and plants. Paper cuts can be valuable presents to show people's happiness and joy.

Paper cutting is one of the most popular and participatory folk arts of China. Ancient Chinese people began to make hollow carved ornaments with gold, silver, leather, or silk fabrics during the Shang Dynasty; this can be seen as the embryonic form of paper cuts.[24] Since the invention of paper by Cai Lun in the Eastern Han Dynasty two thousand years ago, paper cutting has become a popular art using a variety of materials. Paper cuts often appear in religious ceremonies, plastic arts, and in decoration.

Chinese folk paper cutting bears distinct national and geographical features. In artistic style, it is more straightforward in northern China but more exquisite and delicate in the south. The peasant women folk artists use only a pair of scissors and a piece of paper to convey conceptual figures and color effect, producing superb pieces of art. Male artists make paper cuttings with superb delicacy and graciousness in a more realistic style. A typical example is the character 寿 (*shou*), which represents longevity and

31.1 The characters for spring and luck are combined for this Spring Festival decoration.

brings delight to any celebration; and a big red paper character 喜喜 (*xi*) (double happiness) is a tradition on newlyweds' door. During Spring Festival, people paste paper cuttings on windows, door lintels, or use them as table decorations to create a festive atmosphere.

Paper cuts can be produced in a single color and in several colors. Single-color paper cuts are the most common. Such work may seem very simple, but it has definite features. Paper cuts can be multicolored, but this is not common.

Paper cutting was traditionally all done by hand. It is easy to learn the rudiments. Non-artisans need only a knife or a pair of scissors and a piece of paper. Artisans need knives and gravers of various types to make complicated patterns. The paper cut can be one piece of paper or many pieces. Simple patterns can be cut with a knife. To make complicated patterns, people first paste the pattern on the paper and then use various kinds of knives to cut it. No mistakes can be made during the process; otherwise, the work will be spoiled.

31.2 This paper cutting includes the god of longevity and a deity of protection.

There are several steps for the artist to follow when paper cutting. When the artist has the idea for the cutting, she or he must draw it clearly and specifically to have an image in black and white. Then, the artist begins to cut or engrave. After that, she or he removes the cutting carefully and gently, one sheet after another. Finally, the artist copies and pastes to repeat the work, sometimes with improvements.

31.3 The combination of these four characters is displayed to create prosperity.

It is easy to learn about cutting a piece of paper but very difficult to master it. One must hold the knife upright and press evenly on the paper with some pressure; flexibility is required, but wiggling will lead to imprecision or will damage the image. Paper cutters emphasize several styles of line cutting, and there are basic lines that they endeavor to master. They attempt to carve circles like the moon, a straight line like a stalk of wheat, square like a tile and jaggedly, like a man's beard.

In many parts of China, the art of paper cutting has become a requirement for men and women, old and young, and a symbol of a clever mind and nimble fingers. It has always been popular. Nowadays, paper cutting is seen in modern designs such as in product packaging, trademark advertising, interior decorating, fashion design, cartoons, in magazines, or in TV series and on stage as part of the backdrop. As a transmitter of authentic Chinese culture, the art of paper cutting has been introduced to other lands, become the cultural wealth of humanity, and an art treasure for the world.

The oldest surviving paper cut is a round circle from the sixth century, found in Xinjiang Province. By the eighth or ninth century, paper cutting appeared in West Asia, and in Turkey in the sixteenth century. Within a century, paper cutting was done in most of middle Europe. *Scherenschnitte* means "scissors cutting" and is the German and Swiss art of paper cutting. Paper cutting has been a common Jewish art form since the Middle Ages. By about the seventeenth century, paper cutting had become a popular form for creating small religious decorations throughout China. China has many paper cut artists specializing in paper cutouts. The knowledge of this art was usually passed on by generations of paper artists in their family or hometown. There are different styles of rendering paper cuttings in each region of China. There used to be only traditional Chinese themes, but today Western images and modern accents are found in certain paper cutouts.

31.4 There are many Chinese ornaments featuring zodiac animals.

Chinese Culture through Language

剪单页剪纸

Pinyin: Jiǎn dān yè jiǎn zhǐ

Literally: One-page paper cutting.

Meaning: A paper-cutting technique where the artist works with one piece of paper and a pair of scissors.

多页刻纸

Pinyin: Duō yè kè zhǐ

Literally: Multi-page paper engraving.

Meaning: A paper-cutting technique where multiple sheets of paper are cut using a knife to produce several paper cuts.

剪烛西窗

Pinyin: Jiǎn zhú xī chuāng

Literally: Cutting a candle at the window on the west side.

Meaning: Friends and relatives get together and talk intimately.

CHAPTER 32

Chinese Seals

中国印章

The emblem of the 2008 Beijing Olympic Games, "Chinese Seal—Dancing Beijing," made a profound impression on the people everywhere. It showed the world a great deal of historical information and the unique culture of the Chinese seal.

The Chinese seal first appeared in the Spring and Autumn Period. It was a stamp used as a security tool in the process of social communication. The seal began as a simple clay imprint. It was used when ancient people sent or transferred something, in order to prevent it from being opened along the way. They tied the object up with string, then used clay to seal the end of the string, and finally stamped the seal on the clay This kind of mark is the initial seal.

Ancient Chinese seals can mainly be divided into two major categories: official and personal. Seals act as signatures. Possessing the use of an official seal was considered a symbol of power. Both types of seals played a very important role in social and economic interactions. Personal seals are also used for painting inscriptions and are unique works of art in their own right.

In ancient China, there were many kinds of seals with different uses, and they went by many names. After Emperor Qin Shi Huang unified China,

32.1 This seal dates back to the Taiping Rebellion, which lasted from 1850 to 1864.

the seal of an emperor was called *xi*; the seals of officials and civilians were known as *yin*, *ji*, *guanfang*, *hetong*, seal, and *chuozi*, as well as other names, all of which mean seal.

Seal materials are various and have definite grades. For example, *xi* is the emperor's seal, usually made of precious jade. Only emperors, empresses, princes, and marquises could use jade seals and gold seals; high-ranking officials could use silver ones; and low-ranking officials had to use copper. Most official seals are square and little larger than other types. The personal seal is relatively smaller.

To create a seal, one must cut stone and engrave it. In ancient times most of the seals were first traced and then cut. The font on the seals was big seal character, small seal character, and bird script. The bird script is a kind of font made of the decorative shapes of birds and insects. It is considered a special type of calligraphy. The seal character *jing* in the Chinese seal, the emblem of the Beijing Olympic Games, fully demonstrates the art of Chinese seals.

Because the shapes and the inscriptions of seals varied in different periods, people now usually use them for historical accuracy to identify the authenticity of the ancient bureaucratic establishment, geographical features, names, and historical relics. For example, the prints of the private or official seals in the Warring States Period are big seal characters, and those of the Qin Dynasty are small seal characters.

32.2 Many different font styles are used in chop carving.

Chinese seals are also divided into red character seals (*zhuwen* seal), white characters seals, *yangwenyin*, and *yinwenyin*. The characters cut in relief are known as *yangwenyin*, and the characters cut in intaglio (convexly carved) are called *yinwenyin*. In Chinese calligraphy and painting, the characters with vermilion ink paste are red character seals, and the characters carved in intaglio are called white characters seals.

From the Tang Dynasty onward, the seal was no longer a symbol of power or of security. It became a type of art, collected by scholars along with painting and calligraphy.

Seals have remained not only the most important authenticators of official documents or trade contracts but also as an elegant work of art. They are collected and enjoyed by many people because of the seal cutting, drawing, and calligraphy involved in their artisanry.

Seals are commonly used in Asia, especially in China and Japan. At present, the use of seals in Western countries is not as prevalent, as many Westerners simply sign their own names. However, Westerners have also used sealing wax and lute in order to prevent letters and articles from being opened in transit. The ancient Chinese invented lute and sealing wax. Later, Chinese

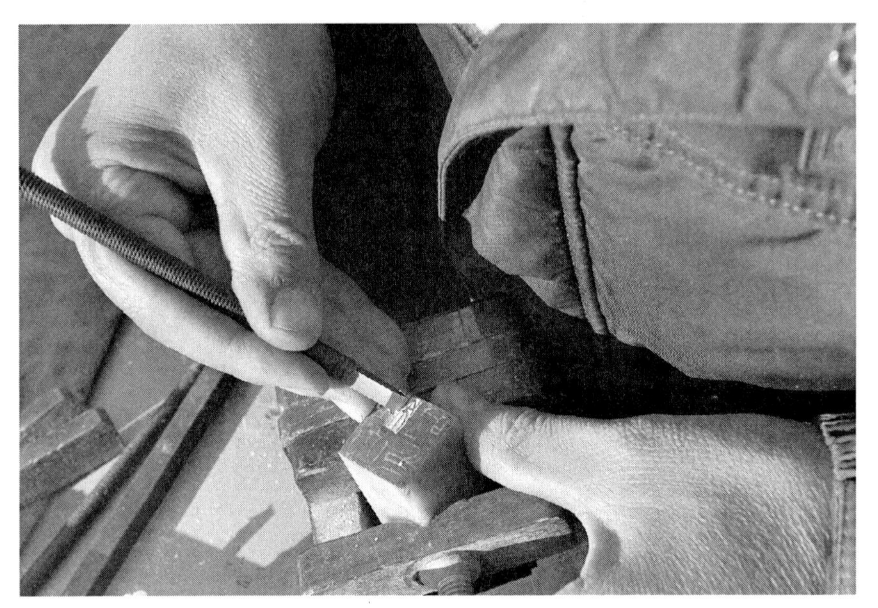

32.3 A seal is laboriously carved into a polished stone face.

sealing wax spread to India, from where it was introduced to Europe and was used to protect secret letters. After putting hot wax onto the envelope flap, Westerners usually imprint their own seal-rings on the wax before it hardens when they are sealing the letters, to prevent other people from opening them. Today, there are still a number of countries and regions using sealing wax in the delivery of certain special items such as cultural relics.

CHAPTER 33

Chinese Musical Instruments

中国乐器

The ancestors of modern Chinese civilization created varied instruments with Chinese characteristics to enrich their lives. Many of these are familiar to Chinese today.

The earliest instrument may be a percussion instrument. According to early records in "The Book of History," (mainly recorded in the Shang and Zhou dynasties) while people were working they found that the dried tree trunks could emit different sounds, so they covered the hollow tree trunks with animal skins to make wooden drums. During the Han Dynasty, drums of various sizes, textures, and effects appeared not only for private entertainment but also for strengthening the military in battle. At present, traditional Chinese musical instruments are divided into wind instruments, bowed instruments, plucked string instruments, and hammered string instruments, which can be summarized as *cui, la, tan, da* (blowing, pulling, plucking, and hammering).

The flute is one of the most common wind instruments. The earliest record of its appearance in the late Spring and Autumn Period. Flutes were popular among the common people and were made from bamboo, wood, iron, and jade. Now most flutes are made of bamboo. Much flute music has been handed down for over the past two thousand years and is very beautiful.

33.1 A transverse flute, also known as a dizi.

Xiao, or the vertical bamboo flute, is an old wind instrument from China and is suitable for solo or ensemble styles of music. In ancient times, *xiao* was made not only of bamboo but also of jade or porcelain. There were only four sound holes at first, and then two were added later. The *xiao* used today has six sound holes and is made of bamboo. Before the Tang Dynasty, there were no differences between the *xiao* and the flute. The instrument played horizontally was known as a flute, and the instrument played vertically was called *xiao*.

The *sheng* (reed pipe), *guan* (pipe), and *suona* (horn) also belong to the family of Chinese wind instruments. *Suona* is a trumpet-shaped instrument, one of the most popular musical instruments of its time. It is said that it came from Persia and Arabia four to five hundred years ago and that *suona* is the transliteration of *surna* (meaning woodwind instrument) in Persian.[25]

It is widely used in local operas and holiday celebrations because of its pitch and clear sound.

The *huqin* (Chinese violin) is the main string instrument in China. There are many kinds of Chinese violin, such as *erhu*, Beijing opera violins, and high-pitched Chinese violins. They are not only suitable for performing deep, plaintive melodies but are also able to demonstrate spectacular momentum; they serve as the main accompaniment in local operas. *Matouqin*, which is well known among Mongolians, also belongs to the string family.

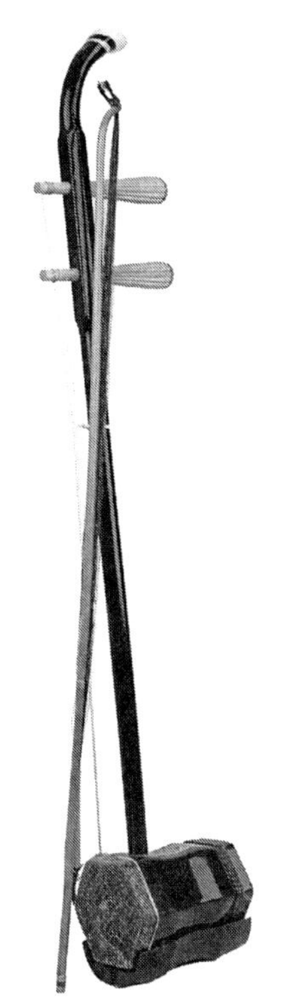

33.2 The erhu is often called the "Chinese violin" or the "Chinese fiddle."

Guqin is the oldest stringed instrument in China. It was very popular in the Zhou Dynasty. The Chinese zither was also popular in the State of Qin during the Spring and Autumn Period and was also known as *qin zheng*. Chinese zithers with twelve to twenty-five strings were usually made of parasol wood. Modern players now like using the twenty-one-string *zheng*. The tone of a *zheng* is vigorous and lingers, making it good at expressing the artistic imagery of floating clouds, flowing water, and general sentiment.

The *pipa* (Chinese lute) was the most common string instrument played in ancient China. According to legend, it was brought back from the Western Regions during the reign of Emperor Wu and became very popular in the Northern and Southern Dynasties. *Sanxian*, also known as *xian zi*, is also a kind of stringed instrument that was played by the Han, the Mongolians, and other minorities.

In addition to drums, the main traditional Chinese percussion instruments are the following. Bell chimes refer to the series of bronze bells hanging on a wooden frame, played with wooden sticks. Since the Shang Dynasty, the bells have been mostly in groups of three. The largest unearthed group has 64 bells.

Generally, chime stones were made of stone or jade and hung on racks. At first they were played for singing and dancing and later were used in sacrificial and other ceremonial activities, becoming a ritual instrument.

33.3 An erhu, pipa, guzheng, and other instruments are included in this ensemble.

The gong is one of the most widely used percussion instruments in China. It is said that it was brought to China from Central and South Asia in the sixth century. Through continuous improvement, the gong can now be played expressively and has become an indispensable instrument in local operas, folk entertainment, and celebrations.

There are also many instruments with the characteristics of Chinese ethnic minorities such as the tamboura (stringed, lute-like instrument), and bongo (hand drum) of Xinjiang Province; the Miao people's *lusheng* (a reed-type wind instrument), bronze drums, and bamboo flute; and the Dai people's *mang gong* and elephant-foot drum (both percussion instruments).

Chinese Culture through Language

对牛弹琴

Pinyin: Duì niú tán qín

Literally: Playing the lute to a cow.

Meaning: Playing to the wrong audience; talking to a wall.

滥竽充数

Pinyin: Làn yú chōng shù

Literally: Make up the number of an orchestra by adding a pretend player.

Meaning: Describes substandard goods that are mixed with quality ones; put in men who cannot work.

老调重弹

Pinyin: Lǎo diào chóng dàn

Literally: Play the same old tunes.

Meaning: To sing the same old song, to say or do the same thing repeatedly.

CHAPTER 34

Peking Opera

京剧

Peking opera is popular throughout the world and is regarded as one of China's national treasures. It is of great influence and an important medium through which traditional Chinese culture is spread. Peking opera is an art form in which a variety of artistic elements are used as symbols of traditional Chinese culture.

Peking opera is regarded as one of China's cultural treasures. At the beginning of the late eighteenth century, music and performance skills were absorbed from the best *Kun* operas (a local drama in Jiangsu Province) and *Qin* operas (Shaanxi Provincial local opera) and were eventually developed into a unique performance style. Talented artists have created and developed opera for two hundred years. It is a subtle and harmonious combination of the various elements of traditional music, poetry, singing, recitation, dance, acrobatics, and martial arts.

The characteristics of Peking opera consist of singing and dancing, which tell a story with singing as monologue and dancing as acts. The dialogue in Peking opera is based on the Beijing dialect, and the songs have a unique pronunciation.

Peking opera employs heavy use of imagery and allegory. The actions in the opera are designed to beautify the scene. In addition to singing, performers

34.1 This warrior costume incorporates long pheasant feathers, which draw attention to the actor's movements.

show their emotions through skillful movements, such as stroking the beard, adjusting the cap, waving with the sleeves and lifting the knees, and raising the feet. For example, the performer shaking the hand and body shows extreme anger; actors put their hands on their heads or throw out their long sleeves to indicate surprise; actors reveal embarrassment by covering their faces with their sleeves.

Some movements in Peking opera are not easy to understand. For example, actors grab their sleeves and then quickly put their hands behind their back in a fast movement, which shows a person is decisive, strong, and is courageous in the face of danger. Sometimes, an action can be done for a long time, as long as twenty minutes. Martial arts fight scenes in Peking opera are a wonderful combination of dance and *wushu*.

In Peking opera, the different roles represent different kinds of people in society, such as the loyal, the cunning, the good and beautiful, and the ugly. The makeup and costume of each role focuses on showing the persona of the character, such as the age, personality, and gender.

34.2

In general, the opera has four roles, *sheng*, *dan*, *hualian,* and *chou*, which are decided by the age and the occupation of the character. *Sheng* are male roles. They can be divided into three categories: the old, the young, and the character good at martial arts. *Dan* are women's roles, including *hua dan, qing yi, wu dan*, and *lao dan*.

Jing, also called *hualian*, is regarded as frank and cheerful, with a bright colored mask. *Chou,* or the clown, is marked with white paint on the nose. Clowns are sometimes positive, warm-hearted, and humorous characters, but at other times they represent dark, scheming, vicious, and stupid people. The singing and performing of the various roles in Peking opera are specialized and have specific skills.

The facial makeup in Peking opera is uniquely Chinese: Each historical role or type of person has a certain kind of makeup called *lianpu*. It is said that this specialized facial makeup in Peking opera was originally a mask. The color of *lianpu* is defined by the role. The types of facial makeup in Peking opera are based on personality, temperament, and other factors. Some colors are reserved for special types of figures. Red, yellow, white,

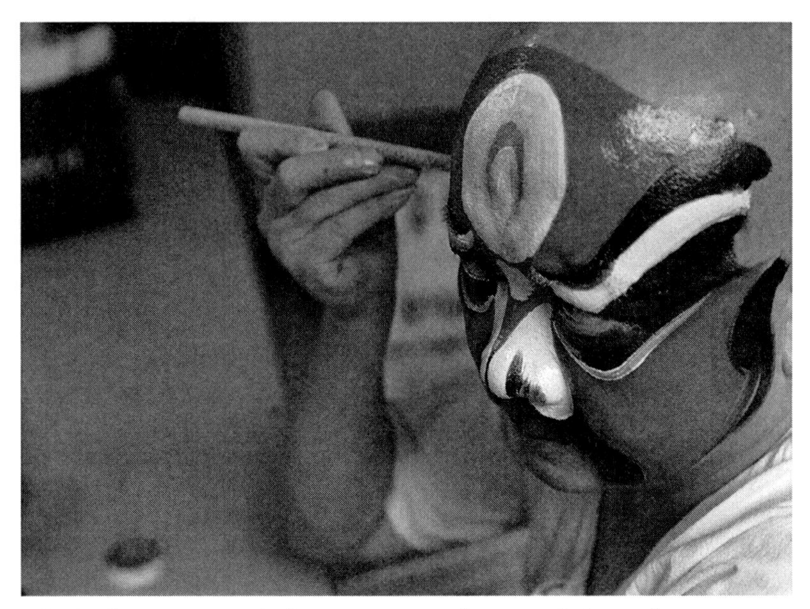

34.3 Performers must spend a long time practicing how to apply their makeup.

black, purple, green, and silver are the main colors used for facial designs to represent different characters. For instance, red represents loyal, courageous, and upright people; white is for sinister and cunning officials; blue shows a character is vigorous, courageous, and enterprising, and yellow shows a resourceful but not extroverted person. Black tells the audience it is a pragmatic role, brown is usually for a stubborn person, and gold and silver stand for gods and ghosts.

The costumes in Peking opera have the characteristics of clothing from the Ming Dynasty. The props, such as cloth walls, tents, capes, whips, paddles, and weapons are also made in the Ming style. The props look real but could not be used in everyday life; however, this does not lessen their power of expression. Exaggeration and symbolism are characteristics of Peking opera. For instance, an actor holding a whip shows that he is riding a fast horse; several soldiers appear on the stage, meaning a complete army. When an actor on stage moves back and forth, it means he is going on a long journey.

34.4 A variety of Peking opera masks.

According to statistics, there are more than 1,300 plays in Peking opera, and there have been many famous Peking opera artists. The well-known actors of the *xiaosheng* role are Tan Xinpei, Yang Xiaolou, Yu Shuyan, Ma Lianliang, and Zhou Xinfang. The most famous actors of *dan* roles are Mei Lanfang, Shang Xiaoyun, Cheng Yanqiu, and Xun Huisheng. Peking opera has two hundred years of history and has become a profound artistic representative of China. It also holds an important position as a performing art globally. In 1919, Mei Lanfang performed in Japan for the first time. Since then, the opera has continued to spread abroad to promote international exchange and the performing arts of China.

Peking opera, a synthesis of music, dance, and acrobatics, is known as China's national opera and widely regarded as the highest expression of Chinese culture. It is one of the most refined forms of opera in the world. Chinese opera distinguishes itself with its rich forms of expression. The events and stories of the performance on stage are reflected not only through music, singing, and actions but also through the masks, costumes, pantomime, dance, acrobatics, martial arts, and sometimes local traditions. The symbols permeate all aspects of Chinese opera and achieve precise characterizations by combining symbols. Gestures, facial expressions and movements characterize actors' behaviors and demonstrate their relationships with each other. Unlike Western operas and musicals, instrumental accompaniment of the Chinese opera is sparse. The two main instruments are the *suona* and the *jinghu*, the latter of which could be described as both a string instrument and a versatile percussion instrument. The difference between Chinese opera arias and recitatives, however, is comparable to those of Western opera.

CHAPTER 35

Sichuan Opera: Face Changing
川剧变脸

At present, there are more than three hundred operas performed in China, among which many of the performances have unique skills. The most magic is the "face changing" of Sichuan opera; its soul-stirring power of artistic expression tends to astound viewers.

Sichuan opera is popular in Sichuan, Chongqing, Yunnan, Guizhou, and Hubei; it is the greatest of the local operas in southwest China. Anyone who has seen it will remember "face changing." Face changing is one type of Chinese opera performance used to show sentiment. The performer expresses sudden changes in mood, such as fear, despair, or anger by changing the color of the face. It was used for characters such as gods and ghosts during the Ming Dynasty. Actors went off stage to change their makeup at that time. Later, actors developed the skill and changed faces on the spot; the technique was spread to acrobatic troupes and other local operas, but Sichuan opera remains the best.

Face changing is an artistic feature in Sichuan opera, shaping the character to reveal the inner thoughts and feelings in a beautiful way. It was said that face changing was adapted by people to fight against ferocious beasts in ancient times in order to survive, and that they outlined different forms on the face in different ways to frighten the beasts. The skill of face changing

35.1

has moved onto the stage and evolved into a unique art. There are two ways to change faces, a total change and partial change. There are three changes, five changes, and nine transitions. Total face changing is divided into three types: wiping the face, blowing the face, and pulling the face. In addition, there is a kind of *qigong* change of face.

Wiping the face: For an actor to paint a particular part of his face, he changes his makeup by wiping his face with his hand while simultaneously performing. If he wants the whole face to change, then paint is applied on the forehead or eyebrows. If he only wants to change the lower half of the face, then the paint can be painted on the face or on the nose; in this way the performer can change his face easily. If the performer only needs to change a certain part of the face, then only a certain position is painted.

Blowing the face: This kind of skill is only suitable for powdered makeup, such as gold powder, graphite powder, and silver powder. A small box of various powders is placed on the floor of the stage. The actor moves with dance movements towards the box, puts his face close to the box, and blows into it. Immediately, the powder from the box changes his face to a different color. Another way is to hide the powder needed in the palm of the actor's hand, which is done by *Xiaoqing*, the character who changes her face several times in the play "The Broken Bridge." The third method is to hide the powder in a wine glass. The actor blows into the glass, and his face changes while he is drinking. When blowing, the performers should make sure not to blow unless the eyes, and mouth are closed, and they hold their breath.

Pulling the face: This is a more complex method of face changing. Actors put the painted *lianpu* on cut pieces of silk, with a thread attached to each mask, and then put them on their face one after another. The silk thread is tied in a certain position of the performer's clothes where it is easy for him to reach without being noticed (such as a belt). As the story progresses, the actors tear the masks off one by one with the dance movements. For instance, the actor could change his face to green, red, white, black and seven other faces in the play "The White Snake." Pulling the face is quite difficult. The glue for the masks should not be too sticky, or it will not be easy to pull the masks off in time, or the masks could all be ripped off at same time. Moreover, the action should be quick and exact, and the actor should distract the audience's attention so that he can finish the change without the audience noticing.

Face changing requires agile movements, which change closely with the mood of the character without the audience noticing. In the development of improving the skills of face changing, the famous Sichuan opera performers Kangzi Lin, Wei Xianglin, Sun Decai, and others have made great contributions.

In short, face changing is one of the unique skills of Sichuan opera, which has now been learned and copied by other local operas. It has even spread overseas. Drama lovers around the world admire it and show great interest in face changing. Many have become keenly interested in Chinese drama, fostering cultural exchange and more interest in China.

Greek tragicomedy, Indian Sanskrit drama, and Chinese opera are indisputably the three most ancient forms of drama in the world. In China, opera has traditionally been the main source of theatrical drama; it is the

35.2 Different faces of Sichuan opera.

Chinese counterpart to Western theater, as in Shakespearean theater. Yet the English language notion of opera does not fully capture the true essence of opera in China, as anyone even vaguely familiar with Chinese opera would agree. In fact, the term "Chinese opera" is not a bad starting point for talking about the phenomenon of opera in China, because the term suggests—and rightly so—that there is something about opera in China that sets it apart from the Western notion of opera.

Face changing is the highlight of Sichuan Opera. It is said that ancient people painted their faces to drive away wild animals. Sichuan opera incorporated this ancient skill and perfected it into an art.

The practice of using bold makeup for different characters involves not only facial makeup but also headdress and specific costumes to fit the role of the character. The painted face no doubt originated as a device to disguise a male actor as a female, providing important symbolic information to the audience. The step from painted face to the idea of changing masks is small, but it has truly made a unique contribution to the art of face changing.

Today, high technology such as lasers and twinkling lights are used to enhance the traditional arts and add a touch of mystery to them.

CHAPTER 36

Acrobatics

杂技

Chinese acrobatics is one of the oldest performing arts. The content is rich and varied; it mainly includes acrobatics, magic, and animal shows. Chinese acrobatic arts are popular with audiences in China and abroad for their colorful and simple style.

Chinese acrobatics appeared in the Spring and Autumn Period 2,500 years ago and were divided into several kinds: strength skills, juggling, juggling with the feet, driving skills, and others.[26] It is said that the earliest acrobatics came from the battle experience of Chinese people. Strength skills is one of the most basic and was developed first. The ancient strong men could easily throw and catch heavy cartwheels and lift big metal cauldrons weighing hundreds of pounds, showing that they not only had strong muscles but had also mastered certain skills.

Chinese acrobatics also evolved from daily life. Performers often made use of common items to show their skills. This includes items such as bowls, plates, cylinders like tubes and pipes, ropes, tables, benches, and bicycles. For example, in a traditional acrobatics program, juggling with the feet, the performers can juggle many common objects, such as wine jars, water barrels, tables, and umbrellas. It is really wonderful when actors sing while they are juggling.

36.1 Demonstrating feats of balance and strength during a Chinese New Year performance.

Acrobatics became a complete art form in the Han Dynasty and reached a wide audience and a high level of mastery. According to Sima Qian's *Shi J* (*Records o the Grand* Histproan) in the spring of 108 BCE, the Han emperor held a grand banquet in the capital, Chang'an, in order to entertain the envoys of Western Region States. During this long banquet he held an unprecedented hundreds of acrobatic shows. Many acrobatic troupes from abroad also took part in the shows.

The acrobatic arts of the Song Dynasty were a turning point for the practice, moving the art away from the palace and into the lives of ordinary people. The actors had to make a living as performers in the street because of losing their original position in the palace. They created new programs, such as kicking skills, illusions, and the skill of taming insects.

During the Qing Dynasty, the content of the acrobatics became richer, and performers' skills improved greatly. Acrobatics could be performed at many different venues: theaters, open-air locations, home celebrations, country markets, or temple fairs. One can still see these wonderful programs from that time, such as playing forks, dancing lions, playing jars, and others.

Some performers also consider magic as a kind of acrobatics. Chinese magic pays great attention to personal skills. A lot of traditional Chinese magic is often based on myths and legends. The performers like to create wonderful fantasy scenes that may incorporate movement and acrobatics.

Chinese acrobatics is undergoing changes. Many cruel and horrible programs such as sword swallowing and lying on a board of nails have been eliminated. A large number of excellent programs, such as diabolo (a juggling

36.2 Foot juggling can be performed with parasols, wooden blocks, fans, and other objects.

36.3 These young women spin three plates with each hand.

prop) juggling, bowl balancing, plate spinning, wire walking, and tightrope walking are still popular. Every year, a lot of acrobatic troupes travel abroad to participate in performances or international acrobatics competitions and obtain many outstanding awards.

Many people think that there are major differences between Western acrobatics and Chinese acrobatics. In the authors' opinion, Chinese acrobatics is known for its breath-taking skills, and Western acrobatics for its ingenious plot land humor, interactive performances which move audiences. But there are many things they have in common, and both have a long history in juggling. For example, in the Spring and Autumn Period, Chinese began juggling with balls and swords, and in an ancient Egyptian tomb of an unknown prince, the murals show circus performers juggling with balls. Now Western and Chinese acrobatics learn from each other and even mix the two styles.

Chinese Culture through Language

踩高跷

Pinyin: cǎi gāo qiào

Literally: Step-walk on stilts.

Meaning: Stilt walking.

走江湖

Pinyin: Zǒu jiāng hú

Literally: Wandering the rivers and lakes.

Meaning: To wander from place to place and earn a living through one's performance, such as juggling, fortune-telling, etc.

叠罗汉

Pinyin: Dié luó hàn

Meaning: Human pyramid: A kind of acrobatic performance where a group of people pile on top of each other in various arrangements to create a pyramid.

竖蜻蜓

Pinyin: Shù qīng tíng

Meaning: A handstand.

走钢丝

Pinyin: Zǒu gāng sī

Meaning: Tightrope walking: this is an acrobatic performance in which an actor or actress performs tricks while walking to and fro on a suspended steel wire.

CHAPTER 37

Crosstalk

相声

Xiangsheng, translated as crosstalk, is the most popular of the ten quyi art forms (unique forms of Chinese performance art). There are three kinds of xiangsheng, dankou, duikou, and qunkou, all of which are conveyed in vivid and humorous language sometimes with singing and dancing performances. It originated in Beijing and has spread nationwide.

Xiangsheng, crosstalk, is one of China's foremost performing arts and a *quyi* art (traditional art), which is comedic and made up of speaking, imitating, teasing, and singing. The language, rich in puns and allusions, is spoken in a rapid, bantering style. It used to be performed in Beijing dialect but is now mainly performed in *putonghua* and sometimes in local dialects. It is very popular throughout the country. Many non-Chinese people have come to China to learn it.

The earliest *xiangsheng* was developed from folk ventriloquism. It emerged in the Qing Dynasty during the reign of Emperor Guangxu as a performance art. Zhu Shaowen, a Peking opera performer, whose stage name was Qing Bupa (meaning Not Afraid of Poverty), used to perform in the entertainment quarter of Beijing known as Tianqiao. The imperial court prohibited performing because of a series of deaths in the royal family, including Empress Cixi's (people were expected to observe a period of mourning). He always began his talk with a ragged verse, imitated street hawkers' peculiar cries, and sang old songs with a pair of bamboo clappers, a fan, or a handkerchief, setting up a piece of cloth as a tent, or simply drawing a circle with white chalk in the street. He later took on two apprentices, and they performed together, so it became crosstalk with a *dougen* and *penggen* (the lead and the stooge).

Modern *xiangsheng* performers must have four skills: speaking, imitating, teasing, and singing.[27] "Speaking" refers to storytelling as well as speech and the way of speaking. "Imitating" means the imitation of a variety of characters or people, dialects, and other sounds, learning to sing famous operas, songs, and dance. "Teasing" means to create humor. "Singing" refers to singing a kind of folk song referred to as "Taiping *geci*," a kind of lyrical oration.

There are three forms of *xiangsheng*: *dankou xiangsheng* is performed by one-person, *duikou xiangsheng* or crosstalk is performed by two people, and the third form, performed by three or more people, is called *qunkou xiangsheng*. In crosstalk, one man is called *dougen*, and the other is the *penggen*. When one is the primary talker while the other chimes in, this is called *yitouchen* (heavy-at-one-end), and the subject of argument between them is called *zimugen*. In *qunkou* one artist must say funny things while the others chime in and make them stray from the subject. Of the three forms, two-person crosstalk is the most popular and widespread.

Xiangsheng is considered the art of laughter. Laughter is the aesthetic principle of crosstalk. Edutainment (education and entertainment) is the art of the comic dialogue, which is the most important feature of the art.

The process of crosstalk performance and appreciation, and the communication between performers and audience, is bi-directional. This characteristic makes it a unique art form, the form of traditional Chinese comedic dialogue. It involves audience participation, which results in unique artistic charm. The relationship between the audience and *xiangsheng* is so close that there is no misunderstanding. *Xiangsheng* draws wisdom and humor from people, playing on their unyielding pursuit for truth and their optimistic spirit, while being satirical towards evil. It has become a brilliant art form known for its exquisite liveliness and its uniqueness.

37.1 The famous xiansheng performer Hu Baolin.

The development of crosstalk has its own characteristics, originating from all walks of the working class. Its main function is to make people laugh, relax, allow them to speak their minds, and express their attitudes. With a history of more than one hundred years, *xiangsheng* has developed into a unique form and a set model for performances from *dankou* and *duikou*, to *qunkou*.

Crosstalk is taught from master to disciple. In the world of crosstalk lineage is important; each student must formally acknowledge someone as their teacher to be classified in a particular school. Otherwise no matter how famous a crosstalk performer is, he or she will always be considered an amateur. There are approximately ten generations of crosstalk performers. Famous first generation performers include Zhu Shaowen and Zhang Sanlu. Ma Sanli, Hou Baolin, Ma Ji, Tang Jiezhong, Jiang Kun, and Zhao Yan are acclaimed performers from the fifth to eighth generations. Due to its humor, sarcasm and unique aesthetic effects, *xiangsheng* has spread far and wide.

Chinese Culture through Language

借古讽今

Pinyin: Jiè gŭ fĕng jīn

Literally: Use the past to satire the present.

Meaning: To use the past to satirize or deride the present.

调侃儿

Pinyin: Tiáo kăn er

Literally: Tease, ridicule, joke.

Meaning: to tease; to joke.

笑破肚皮

Pinyin: Xiào pò dù pí

Literally: Laugh and break the skin of your stomach.

Meaning: to split one's sides with laughter.

说唱文学

Pinyin: Shuō chàng wén xué

Literally: Speak-and-sing literature.

Meaning: Narrative art involving speaking and singing.

CHAPTER 38

Auspicious Culture
吉祥文化

Auspicious culture has played an important role throughout the development of Chinese traditional culture. It condenses Chinese people's moral feelings, life awareness, aesthetic, and religious sentiments. Its aim is to help people and inspire them to create a better life. The thoughts of Chinese people are reflected in auspicious culture, and it is composed of the eternal theme of culture, auspiciousness, and beautiful imagery. Auspicious culture has influenced people from ancient times to the present day and still exists everywhere in China.

In the development of China's five thousand years of history, people have expressed their wishes for a happy life by making use of Chinese characters, beasts, flowers, birds, heavenly bodies, natural phenomena—such as thunder and lightening—by way of metonymy, personification, puns, homonyms, and symbols; it is from these that Chinese auspicious culture has been formed.

Auspiciousness means showing or suggesting future success is likely. Good luck means a force that brings good fortune; thus, everything is satisfactory and pleasant. Observers can see the life consciousness, the appealing esthetic, and national qualities of Chinese people through decorative designs, patterns, and signs. There are many things that carry good luck. The most common are explained here.

38.1 Fish are a symbol of abundance and unity.

Lucky animals

Chinese people think animals can bring them good luck. Some animals that embody good fortune exist in the real world, and some are imaginary. The unicorn, the phoenix, the turtle, and the dragon were called The Four Spirits by the ancient Chinese people. These are clever and lucky animals, but only turtles exist in the real world. The turtle symbolizes longevity because it has a long life. The unicorn, the phoenix, and the dragon symbolize peace, nobleness, and bravery respectively, and are mythical animals.

In Chinese culture, the dragon and the phoenix are the most popular lucky symbols. You can see them everywhere, especially on traditional bridal dresses, where patterns of a dragon and phoenix are embroidered.

Auspicious plants and food

Flowers, trees, and fruit carry auspicious meanings. The term *The Three Friends in Winter* is the pairing of plum blossoms, bamboo, and pine trees to express auspicious content. When the cold of winter comes they do not wither and die like other plants and have come to symbolize perseverance, integrity, and modesty. Red beans convey the idea of thinking of somebody. It is a traditional habit for lovers to present red beans to each other as gifts. Orange, for instance, in Chinese 桔 (jú), is a homonym of 吉 (jí, lucky). People wish good luck to each other through presenting oranges. The pomegranate is a symbol of many children and happiness. The orchid is famous for its unique, delicate fragrance, and it is regarded as the nobility of flowers, a symbol of elegance. The peony is a combination of fragrance, color, and elegance.

Auspicious numbers

In Chinese culture, numerals not only tell quantities but carry auspicious implications. "One," "ten," "hundred," and "thousand" together with several other numerals bring luck. Both even and odd numerals have propitious meanings, and they have always played a very important role in the history of Chinese culture. It is said there were once 9,999 rooms in the Forbidden City! (Nine is a very lucky number.) The most popular lucky numbers are three, six, eight, and nine. Chinese people like to have these numbers in their telephone numbers, license plates, or room numbers because they think these numbers can bring them good fortune.

Characters

Chinese characters themselves are great decorations. Therefore, people love to make lucky writings in various forms of calligraphy. The most commonly used four characters are "fortune," "wealth," "longevity," and "happiness." Fortune (福) means good luck and blessings, and wealth (财) means success in career. Longevity (寿) represents a long life, and happiness (幸) symbolizes enjoyment and delight. Chinese people like to decorate all kinds of things with many designs of these characters to express their good wishes.

38.2 The Chinese character for good luck and blessings is displayed in many homes and businesses.

Auspicious language

China is a nation of etiquette. For example, Chinese people say "I wish you a prosperous new year" during Spring Festival and wish everyone they meet prosperity and wealth in the coming year. When attending a wedding ceremony, one must use the set phrase to address the newlyweds, "May you live together all your life." When elderly people have a birthday, people must say "Be as deeply happy as the South Sea and live as long as the mountains" to bless them with a long and a happy life. When a store opens for business, Chinese say something to show their wish that business will get better and better. Many different words are used to wish that "May everything go well," "May good wishes come true," or "Make progress in your studies," all in an effort to express good wishes in daily life.

Auspicious behavior

In China, people like to do anything lucky to express their good wishes. The most typical time to do this is during Spring Festival (New Year). Chinese people put couplets or the character *fu* (福, luck) on the door to stimulate the festive atmosphere to show their wishes. All family members have a rich dinner (*nianfan*) together to show they are united; the older members offer money in red pockets to the young people, blessing the children and wishing them health in the coming year. Setting off firecrackers indicates welcoming the new year and scaring away the old spirits; eating *jiaozi* (dumplings) and *niangao* (sticky rice cake) symbolizes good fortune and wealth. Relatives and friends visit each other to bestow good wishes on each other.

It is obvious that Chinese lucky culture can be found in all parts of Chinese everyday life.

There are different auspicious cultures in different countries. For instance, in Chinese culture, certain numbers are believed to be auspicious or inauspicious. Lucky numbers are based on Chinese words that sound similar to other Chinese words. The numbers six, eight, and nine are believed to have auspicious meanings because they sound similar to words that have positive meanings. Number four (四, *sì*) is considered unlucky in Chinese, Korean, Vietnamese, and Japanese cultures, because it is nearly homophonous to the word "death" (死, *sǐ*). Number fourteen is considered one of the unluckiest numbers. Fourteen is usually pronounced as 十四 (*shí sì*) in *putonghua*, which sounds like 十死 "ten die," but in the West, thirteen is considered unlucky.

38.3 Bats are auspicious because they are a homophone for good fortune and happiness.

However, thirteen has a positive connotation in southern China and Hong Kong, as it sounds like *yat sang* in Cantonese, meaning "being alive." Another interesting example is the color red. Whether in most Western countries or in China, red is related to celebrations and happy days. In China the phrase 开门红 (*kāiménhóng*) is used, which means "good luck" or to be off to a good start. This is slightly similar to the English phrase "red-letter day," which connotes a something of special significance or importance will happen. In the West, the owl is wise, clever, and just; hence, the English saying "as wise as an owl." But in Chinese, owl means "bad luck." Many Chinese people are even afraid of their hoots.

Chinese Culture through Language

世上无难事，只怕有心人

Pinyin: Shì shàng wú nán shì, zhǐ pà yǒu xīn rén

Literally: There is nothing difficult in the world for those who put their heart to it.

Meaning: Nothing in the world is difficult for one who sets his mind to it.

寿比南山，福如东海

Pinyin: Shòu bǐ nán shān, fú rú dōng hǎi

Literally: Longevity like the South Mountain, good fortune like the East Sea.

Meaning: To wish someone a long and happy life.

金玉满堂

Pinyin: Jīn yù mǎn táng

Literally: Gold and jade fill the hall.

Meaning: May you have many children, may your house be full of treasures.

吉人天相

Pinyin: Jí rén tiān xiàng

Literally: Heaven helps a good man.

Meaning: Knock on wood; be a good person.

龙凤呈祥

Pinyin: Lóng fèng chéng xiáng

Literally: Prosperity brought by the dragon and the phoenix.

Meaning: In extremely good fortune; twin bliss.

生意兴隆

Pinyin: Shēng yì xīng lóng

Literally: Business flourishes.

Meaning: May your business prosper.

吉祥如意

Pinyin: Jí xiáng rú yì

Literally: Auspicious as desired.

Meaning: Everything goes well; be as lucky as desired.

招财进宝

Pinyin: Zhāo cái jìn bǎo

Literally: Bring in wealth and treasure.

Meaning: Felicitous wish of making money.

一帆风顺

Pinyin: Yī fān fēng shùn

Literally: One sail, smooth wind.

Meaning: Wishing you every success; smooth sailing.

步步高升

Pinyin: Bù bù gāo shēng

Literally: Move up step by step.

Meaning: To be promoted to a higher position.

CHAPTER 39

Cultural Taboos

忌讳文化

Taboos reflect the nature of people and social relationships. Since ancient times, Chinese people have observed and recorded many taboos in personal and public writings. Taboos play a very important role in our life and behavior and influence people both physically and spiritually in many ways.

Taboos are the products of society and culture: They had appeared a long time ago in Chinese history. Generally, taboos are something people avoid doing or saying because society thinks it is offensive, embarrassing, or wrong. Taboos exist in every place of life and affect our daily life.

In ancient times, the names of the emperor and Confucius were not allowed to be spoken or written directly. It was a national taboo. It is a family taboo to say the names of the family ancestors (alive or dead) out loud. There are many historical examples of such taboos: The emperor was named Ying Zheng (he took the thrown in 246 BC and was later known as Qin Shi Huang, the first emperor), so the pronunciation of the first month of the year was changed to *zhengyue*. Sima Qian's book was titled *Shiji* (*ji* meaning record) instead of *Shitan* (*tan* meaning discussion), because his father's name was Sima Tan; if his book had the same name as his father, then each time he said the name of the book it would sound as if he was speaking his father's name. By the Tang Dynasty, in order to respect ancestors and maintain authority at home, people were not allowed to use the names of the monarchs for more than seven generations, even after they died. Mirrors were considered taboo in the Song Dynasty, because the emperor's great grandfather's name was Jing, the same sound as the word for mirror.

There were three ways to avoid taboos in ancient times. First, replace the character for the name of the monarch or the ancestor with other characters. Second, do not write the character, but leave it blank. Third, write the character incorrectly by advertently leaving out the last few or final strokes.

Chinese people believe that numbers are auspicious and ominous. There are many old sayings about numbers: "even means good" and "good luck begins from eight." When numbers are considered taboo, certain ones must be avoided, as many people feel they represent danger, loneliness, departure, or death. If you give gifts to those who get married or have a birthday, the gift should be of an even number; but if you go to visit a patient or attend a funeral, an even number should be avoided, because nobody wants bad things in pairs. The number three, (三) pronounced *sān* in *putonghua* is homophonic with *sàn* (散) which means "to break up." However, Cantonese like the number "three" because it is homophonic with "life." Another example is that men do not celebrate their thirtieth birthday, but women don't celebrate their fortieth, because three means to "break up" and four means "die." Why can't women celebrate their thirtieth but men can? In earlier times, women had no social position, so they were afraid to "break up" when they celebrated their thirtieth birthday. Men celebrate their fortieth birthday because they believed they would not die.

Han Chinese traditionally avoided getting married in May, July, and September, which were considered inauspicious months. People should not get married at eighteen, for the couple would have eighteen hardships. In the city of Kaifeng, Henan Province, and some other places, people believe elderly people will die either at age seventy-three or eighty-four. These numbers are taboos, of course, and lack scientific justification, but they are expressions of people's loneliness, fear of death, their good wishes for longevity, and human instinct to protect themselves.

In the past there were many taboos involving women; they were not allowed to go out to see neighbors, relatives, or friends in the first month of the New Year. It is traditional that women should not visit others at their houses in the month after they have given birth, for it is said that they would dirty others or they would be unlucky. Widows were considered bad luck. People also avoided nuns (and monks). In Hunan, it was said that if you encountered nuns or monks on New Year's Day, you would be unlucky for a year. In many places it is still considered taboo for those who have had a death in the family to pay visits to others during the Spring Festival the following year.

There are taboos for objects like animals, plants, and artifacts. For instance, Chinese consider tigers to represent mountain spirits or beasts; snakes are long worms. The popular belief is that, if you point at a snake with your fingers, your fingers will become a snakehead. It is also considered bad luck to hear animal sounds, such as the barking of dogs, which sounds like crying or somebody dying nearby. The caw of crows is considered a bad omen throughout the country, and owl hoots at night are also bad luck.

As for plants, old trees beside temples and villages should not be cut. Objects such as umbrellas cannot be pronounced as *san*, because, like three, it is the homophone of "break up." Clocks should never be given as gifts, because "clock" sounds like "the end" (of life). There are twelve taboos for chopsticks. For example, if you insert a chopstick into a bowl of rice at the table, it is considered a humiliation to others. In addition, you should not cross your chopsticks casually on the table, as it reminds people of funeral ceremonies.

39.1 It is taboo to insert chopsticks vertically in a rice bowl as it looks similar to burning incense for the deceased.

People are not allowed to say "disease," "death," "ghost," "poverty," and other unlucky words on Chinese New Year's Eve and Chinese New Year. People should not break any bowls, and benches should not fall over on New Year's Eve; otherwise, something unlucky may happen in the coming year. There is another old saying that if you have your hair cut in January, your uncle will die. In the past, people would never eat porridge or meat for breakfast on New Year's Day. Instead, families must eat cooked rice in the morning on New Year's Day so that the whole family will become very rich in the coming year (in the past, only poor people ate rice porridge). Many people also believe one should not wash clothes on the first or second day of the New Year, because it is the water god's birthday.

There are a variety of other taboos in China. At different times, different areas had different taboos. These may now have changed in some ways, but many traditional Chinese taboos still remain. Some have something to do with superstitions and may not be scientific, whereas others are contribute to daily life and experience, like cleaning your house so that you can start the New Year clean and organized. They play a role in regulating people and society and help people deal with interpersonal relationships.

Taboos and superstitions are cultural phenomena in all nations and regulate people's language and social communication. In Western countries, Christians believe that the misuse of the name of God is blasphemous. This is true for Chinese as well: One must try to avoid saying the name of God or even the names of the ancestors. In the West, people do not like to talk about disease and death, and neither do Chinese. There are dissimilarities too. For instance, many Westerners are superstitious about thirteen, but Chinese do not like fourteen; Westerners think of dragons as cruel and vicious, but Chinese consider dragons noble and auspicious creatures, and so on. In short, understanding the phenomenon of taboos and superstitions in different cultures can help us have deeper cross-cultural communication.

Chinese Culture through Language

入乡随俗
Pinyin: Rù xiāng suí sú
Literally: Enter a village and follow its customs.
Meaning: When in Rome do as the Romans do.

如意算盘
Pinyin: Rú yì suàn pán
Literally: Smug calculation with an abacus.
Meaning: Wishful thinking; smug calculation.

送瘟神
Pinyin: Sòng wēn shén
Literally: Sending the God of Plague away.
Meaning: Get rid of somebody or something unwanted or unpleasant.

本命年
Pinyin: Běn mìng nián
Literally: Zodiac year of birth.
Meaning: It refers to the animal year in which a person is born according to the Chinese zodiac, which occurs every twelve years. There are many superstitions around this year.

CHAPTER 40

Chinese Feng shui
风水文化

In recent years, more and more non-Chinese people have become interested in feng shui. What's more, feng shui has been introduced into the school curriculums in some universities. Now people are studying ancient feng shui in conjunction with modern scientific methods.

There are a lot of definitions of *feng shui* in China. Generally speaking, it is related to the placement of homes and graveyards based on topography and direction. It can also be interpreted as a type of superstition because of the way ancient people connected it with destiny and the good and bad fortune of their future generations. However, the common explanation is that *feng shui* is very useful technology in selecting the correct environment. Ancient people adapted their surroundings to create an auspicious site and layout while adding artistic merit. Many people consider *feng shui* rules when choosing a location, such as the effects of light, airflow, and water, which also coincide with many findings of modern science. Of course the superstitious legends and explanations of *feng shui* do not have much in common with science. Chinese *feng shui* culture contains a lot of useful information about architecture, aesthetics, and ecology.

Feng shui influenced the location of villages in ancient China. A site surrounded by water or by a river and in front of hills was the ideal site for many villages. Other villages were set up in front of or behind water or rivers because the villagers believed the river was their guardian.

40.1 This special compass, called a luo pan, is used specifically to determine the feng shui *of a location.*

Since ancient times, Chinese have paid great attention to choosing a dwelling place. It was, and often still is, believed that *feng shui* affects the prosperity of a family and influences people's physiology, mental state, and even emotions and ethics. Generally, it is best to live in a location facing water and with hills behind it. Also, the gate or door is very important because, according to *feng shui,* it can create harmony and deter harm. Mountains and water are the most auspicious things in nature for Chinese people, so the gate or door must always open towards them.

Although there is some superstition in *feng shui*, people have studied it and gained some scientific knowledge about aspects of ecology, psychology, hygiene, and aesthetics contained within *feng shui*. What is called *sha* or "evil" in Chinese *feng shui* (similar to geomancy) are actually elements that may not be advantageous or even be harmful to a living environment, such as a certain direction, quality of light, sound, and even color. According to the theories of the five natural elements of metal, wood, water, fire, and earth, all things on earth can both promote and restrain each other. In order to keep the balance, *feng shui* is supposed to help people promote harmony and avoid harm or *sha*.

In daily life, people can beautify the environment and adjust their mentality to reach a harmonious state. Chinese culture developed with a belief in the middle path. For example, it may be harmful for people to live in a room that is too bright or too dark, but you can mediate or balance it with *feng shui*.

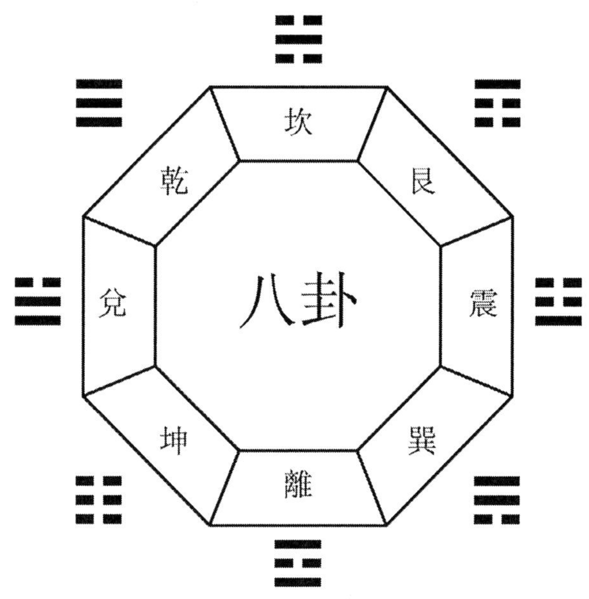

40.2 A feng shui bagua is one of the main tools used to analyze the feng shui of any given space.

Some Chinese people like to keep a fish bowl or place some flowers and other auspicious things in the office to bring good luck in dealing with office politics.[28] Some feng shui masters believe that it helps, because if people have these psychological suggestions and a pleasant environment, they are more relaxed and more confident. As more people learn about *feng shui* over time, it will be better understood.

Feng shui originated in China and is an important part of Chinese traditional culture. Since the 1880s, *feng shui* has spread throughout the world for many reasons, such as concern about environmental pollution, population explosion, resource crises, soil erosion, climate changes, and so on. Often, the relationship between people and the environment is overlooked in modern architecture; however, it is well considered in Chinese *feng shui*. It not only meets the psychological needs of the people but also has a certain practical value. It is reported that biomorphic houses are very popular in Germany, and the environment of the houses is selected carefully according to Chinese *feng shui*. Although many Americans didn't believe in *feng shui* in the past, it has been introduced into the curriculum in many universities.

According to Tokyo University professor Watanabe Kinxiong, there are more than one hundred universities offering *feng shui* courses in Japan.

Chinese Culture through Language

风水宝地
Pinyin: Fēng shuǐ bǎo dì
Literally: Wing-water-treasure-land.
Meaning: A place with good *feng shui*.

风水轮流转
Pinyin: Fēng shuǐ lún liú zhuàn
Literally: The wind and water take their turn.
Meaning: Every dog has his day; the tables have turned now.

五行之说
Pinyin: Wǔ xíng zhī shuō
Literally: The priciples of the five elements.
Meaning: The five elements of metal, wood, water, fire and earth are a key feature of ancient Chinese culture, and a corner stone of traditional chinese medicine.

Image Sources

1.1 bigstock-Chinese-Dragon-2008233
1.2 bigstock-Chinese-Teapot-2751172

2.1 bigstock-mount-huangshan-23887670
2.2 http///commons.wikimedia.org/wiki/File/Mount_tai_rock_inscriptions.jpg
2.4 bigstock-China-Hukou-Waterfall-of-the-Y-30261986

3.1 bigstock-Taoism-In-Yunnan-China-27096710
3.2 http://commons.wikimedia.org/wiki/File-Leshan_Giant_Buddha.jpg
3.3 https-//zh.wikipedia.org/zh-hk/File-Status_of_Kuan_Yin.jpg
3.4 bigstock-Grand-Buddha-Statue-In-Leshan–2256657

5.1 http://www.bigstockphoto.com/image-16566029/
stock-photo-chinese-new-year-ornaments-and-candy-box 16566029
5.2 bigstock-MALACCA–FEBRUARY—Lion-da-12173435
5.3 bigstock-Dragon-Boat-Race-4512699
5.4 bigstock-Chinese-famous-food-Mooncakes-32360519

6.1 Courtesy of Access Travel. www.accesschinatravel.com/
article-Yuyuan-Garden-Map
6.2 http://commons.wikimedia.org/wiki/File:2004_0927-Suzhou_
MasterOfNetGarden_PaintedMap.jpg
6.3 bigstock-Chinese-garden-n-Zurich-26613389

7.1 http://zh.wharugo.com/%E5%9B%9B%E5%90%88%E9%99%A2
7.2 bigstock-Fujian-Tulou-in-China-old-bui-28475747
7.3 http://commons.wikimedia.org/wiki/
File-Cave_Dwelling_-_Courtyard.jpg Cave_Dwelling_-_Courtyard

8.1 http://commons.wikimedia.org/wiki/File:%E7%99%BD%E9%A9%
AC%E5%AF%BA.jpg
8.2 http://commons.wikimedia.org/wiki/File-Guashan_Shaolin_temple
.jpg

9.1 bigstock-Traditional-oriental-lunar-cal-27365999
9.2 bigstock-Chinese-calendar-book-9660314

10.1 Pen_ts'ao,_woodblock_book_1249-ce http://zh-yue.wikipedia.org/
wiki/File-Pen_ts%27ao,_woodblock_book_1249-ce.png.png
10.2 bigstock-firecrackers-5744508
10.3 bigstock-Ancient-Chinese-Compass-On-Whi-5589284

11.1 http://en.wikipedia.org/wiki/File-Shang_dynasty_inscribed_scapula.
jpg Shang_dynasty_inscribed_scapula
11.2 bigstock-Chinese-old-oracle-with-animal-34442090
11.3 http://zh-yue.wikipedia.org/wiki/File-OracleShell.JPG

12.1 http://zh.wikipedia.org/wiki/File-Ink_brush-xiangshan.jpg
Ink_brush-xiangshan
12.2 bigstock-Four-Treasure-Of-The-Study-3953501
12.3 Duanyan http://zh-yue.wikipedia.org/wiki/File-Duanyan.jpg

13.1 Lian_Zhu_Shi http://zh-yue.wikipedia.org/wiki/
File-Lian_Zhu_Shi.jpg
13.2 bigstock-Chinese-Chess–Stand-For-Star-5125742
13.3 bigstock-writing-Chinese-Calligraphy-23089250

14.1 bigstock-Chinese-Market-Goods-1408765
14.2 bigstock-Chinese-doctor-checking-a-pati-6252016
14.3 bigstock-Chinese-Herb-Collection-6876773

15.1 bigstock-Chinese-Elderly-Woman-Performi-18371273
15.2 bigstock-chinese-food-therapy-traditio-31099490

16.1 bigstock-Shaolin-Kung-Fu-4737365
16.2 bigstock-Wudang-Taiji-Quan-13553612

17.1 bigstock-Weapons-762477
17.2 bigstock-Guan-Dao-Kwan-Dao-Chinese-Pol-35204612
17.3 Shaolin-show http://commons.wikimedia.org/wiki/File-Shaolin-show .jpg

19.1 Daoist-symbols_Qingyanggong_Chengdu
19.2 Commons Wikimedia File-PapperCutPig
19.3 http://en.wikipedia.org/wiki/File-20100720_Fukuoka_ Kushida_3614_M.jpg 20100720_Fukuoka_Kushida_3614_M

20.1 bigstock-Chinese-Wedding-Dress-1406527
20.2 bigstock-China-Big-Sedan-Chair-42035545
20.3 Traditional_chinese_wedding https-//zh.wikipedia.org/wiki/ File-Traditional_chinese_wedding.jpg

21.1 AltechinesischeMuenzen https-//zh.wikipedia.org/wiki/ File-AltechinesischeMuenzen.jpg
21.2 bigstock-Chinese-Dragon-coin-34754741
21.3 China_coin1 https-//zh.wikipedia.org/wiki/File-China_coin1.JPG

22.1 bigstock-Mapo-Tofu–A-Popular-Chinese–4280198
22.2 bigstock-Chinese-Sichuan-Cuisine-15375410
22.3 bigstock-Chinese-Cook-Prepares-Peking-R-7444475
22.4 bigstock-Whole-Peking-Duck-8249144
22.5 bigstock-Chinese-Traditional-Style-Kitc-15383051

23.1 Chinesericewine https-//zh.wikipedia.org/wiki/File-Chinesericewine.jpg
23.2 http://zh.wikipedia.org/wiki/File-Rice_Wine.jpg

24.1 bigstock-Chinese-tea-16565240
24.2 bigstock-Set-With-Tea-Leaves-34334276
24.3 bigstock-Traditional-chinese-tea-set-6311685

25.1 bigstock-China–Different-Minorities-3859034
25.2 http://zh.wikipedia.org/wiki/File-%E6%AF%93%E6%9C%97%E8%B4%9D%E5%8B%92%E7%A6%8F%E6%99%8B.jpg
25.3 bigstock-Ancient-Chinese-dress-779983
25.4 http://zh.wikipedia.org/wiki/File-%E3%80%8A%E7%8E%AB%E8%B4%B5%E5%A6%83%E6%98%A5%E8%B4%B5%E4%BA%BA%E8%A1%8C%E4%B9%90%E5%9B%BE%E3%80%8B.jpg

26.1 bigstock-Adult-bound-feet-chinese-shoes-32334680
26.2 http://zh.wikipedia.org/wiki/File-Qifu_late_19th_or_early_20th_century,_Honolulu_Academy_of_Arts

27.1 bigstock-Chinese-Stone-Table-11805833
27.2 bigstock-Chinese-antique-ming-style-fur-26783303
27.3 bigstock-Sofa-1857519

28.1 bigstock-Chinese-Shadow-Figures-3299354
28.2 bigstock-a-string-of-colorful-chinese-k-22371911
28.3 bigstock-LONGJI-CHINA–MAY—Yao-et-12174122

29.1 bigstock-Jade-Kylin–Chinese-Unicorn–13592111
29.2 bigstock-Chinese-famous-gemstone-Shoush-20204330

30.1 bigstock-Chinese-Antique-Vase-23827976
30.2 bigstock-Chinese-Vase-3644818
30.3 bigstock-Chinese-Teapot-2751172
30.4 bigstock-Close-up-Of-Cloisonn–Manufac-4982012

31.1 bigstock-Paper-cutting-isolated-on-whit-20268719
31.2 http://commons.wikimedia.org/wiki/File-Daolingzhi.jpg
31.3 http://commons.wikimedia.org/wiki/File-Zhao_cai_jin_bao_papier_decoupe_chine.jpg
31.4 http://commons.wikimedia.org/wiki/File-Jianzhi.JPG

32.1 http://commons.wikimedia.org/wiki/File-TaiPingRevolutionSeal.png
32.2 bigstock-Carving-A-Chinese-Seal-11334719
32.3 bigstock-Happy-Chinese-stamp-8043032

33.1 bigstock-Chinese-Transverse-Flute-Dizi-3034556
33.2 bigstock-Ehru-Chinese-Violin–1077114
33.3 ConcertGroupPano http://zh-yue.wikipedia.org/wiki/File-ConcertGroupPano.jpg

34.1 bigstock-BEIJING-CHINA–NOVEMBER—24772193
34.2 bigstock-Beijing-Chinese-opera-12165974
34.3 bigstock-Chinese-Opera-Actor-Is-Paintin-18281843
34.4 bigstock-Peking-Opera-653941

35.1 http://zh.wikipedia.org/wiki/File-Bianlian.JPG
35.2 bigstock-Change-Face-2448342

36.1 bigstock-Chinese-New-Year—celebrat-28708424
36.2 bigstock-Great-Chinese-Circus-Confucius-5095256
36.3 bigstock-BEIJING-CHINA–JUNE–Balan-12172091

37.1 Huo_Baolin_1960 http://en.wikipedia.org/wiki/File-Huo_Baolin_1960.jpg

38.1 http://www.bigstockphoto.com/image-27369488/stock-photo-chinese-new-year-ornament-traditional-fabric-fish-symbolizes-prosperity-and-good-luck 27369488
38.2 http://www.bigstockphoto.com/image-12176612/stock-photo-auspicious-symbol-fu-and-chinese-decorative-knots 12176612
38.3 http://www.bigstockphoto.com/image-1677379/stock-vector-chinese-bat-pattern-%28-vector-%29 1677379

39.1 http://www.bigstockphoto.com/image-21029501/stock-photo-funeral-rice

40.1 bigstock-Fengshui-Compass-1601167
40.2 http://commons.wikimedia.org/wiki/File-Acht-trigramme.svg

Endnotes

1. Kaicheng Zhang and Anning Hu, *Dragon Culture—Review and Prospect* (Qingdao: Qingdao Ocean University, 1994), 8.
2. Jianqing Guo, *An Overview of Chinese Culture* (Shanghai: Jiaotong University Press, 2005), 176.
3. Robert Temple, *The Genius of China: 3,000 years of Science, Discovery, & Invention* (Vermont: Inner Traditions, 2007), 1–5.
4. Jiantang Han, *The Culture of Chinese Characters: A Pictoral Exploration* (Beijing: Beijing Language and Culture University Press, 2005), 8.
5. Xiuping Zhang and Naizhang Wang, *Effects of 100 Kinds of Culture in China* (Nanning: Guangxi Peoples Publishing House, 1994), 371.
6. Jianzhong Li, *A Brief Introduction to Chinese Culture* (Wuhan: Wuhan University Press, 2005), 177.
7. Zhang and Wang, *100 Kinds of Culture*, 159.
8. Xiaojiang Zheng, *The Mystery of Chinese Culture* (Beijing: Contemporary World Press, 2008), 146–147.
9. Zhang and Wang, *100 Kinds of Culture*, 422.
10. Li, *Introduction to Chinese Culture*, 247.
11. Chunmiao Zheng, *Comparison between Chinese and Western Culture* (Beijing: Beijing Language College Press, 1994), 109.
12. Zhiming Hu and Xiaoheng Zhang, *Common Sense in Chinese Culture* (Beijing: Chinese Movie Press, 2007), 133.
13. Yifei Wang, *Splendid Traditional Chinese Culture* (Beijing: Beijing Jihui Ltd. Co. Publishing House, 2007), 262.
14. Hu, *Common Sense*, 202.
15. Ibid., 205.
16. Zhaoyan Xia, *Introduction to Chinese Culture* (Haikou: Southern Press, 1999), 436–437.
17. Pixi Chen, *Dress Culture* (Beijing: China Economic Press, 1995), 70.

18. Yuhong Li, *The History of Chinese and Foreign Furniture* (Harbin: Northeast Forestry University Press, 2000),1808.
19. Zhang and Wang, *100 Kinds of Culture*, 560.
20. Haifang Sun, *Yue Kiln Celadon in China* (Shanghai: Shanghai Ancient Books Publishing House, 2007), 1–3.
21. Zhongping Yan, *Chinese Modern Economic History Statistics* (Beijing: Science Press, 1955), 100.
22. Ning Gang and Kong Zhengzhen, "A Comparative Study on the decoration of porcelain between Chinese and western style," *Jiaoxue Lunwen* (September 2011): http://www.11yj.cn/taoyi/jiaoxue/lunwen/08196723.htmlrn
23. Adolf Reichwein, *Cultural Contact between China and Western Countries in the 18th Century* (Beijing: The Commercial Press, 1962), 79–90.
24. Xiuping Zhang and Naizhuang Wang, *An Overview of Chinese Culture* (Beijing: Dongfang Press, 1988), 87.
25. Jinyuan Zhu, *The ABCs of Traditional Culture* (Jinan: Shangdong Friendship Press, 1996), 143.
26. Zhang and Wang, *100 Kind of Culture*, 321–322.
27. Zhang and Wang, *Overview of Chinese Culture*, 364.
28. Xia Li, *50 Topics on Chinese Culture* (Beijing: Foreign Language Press, 2007), 111.